T0269972

HAVE A GOOD TRIP

ALSO BY EUGENIA BONE

Nonfiction

Fantastic Fungi: How Mushrooms Can Heal, Shift Consciousness, and Save the Planet (contributor)

Microbia: A Journey into the Unseen World Around You

Mycophilia: Revelations from the Weird World of Mushrooms

Cookbooks

Fantastic Fungi Community Cookbook

The Kitchen Ecosystem: Integrating Recipes to Create Delicious Meals

Well-Preserved: Recipes and Techniques for Putting Up Small Batches of Seasonal Foods

Italian Family Dining (with Edward Giobbi)

At Mesa's Edge: Cooking and Ranching in Colorado's North Fork Valley

HAVE A GOOD TRIP

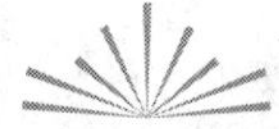

Exploring the Magic Mushroom Experience

EUGENIA BONE

FLATIRON
BOOKS
NEW YORK

Printed in the United States of America. For information, address Flatiron Books, 120 Broadway, New York, NY 10271.

www.flatironbooks.com

Designed by Susan Walsh

Library of Congress Cataloging-in-Publication Data

Names: Bone, Eugenia, author.
Title: Have a good trip : exploring the magic mushroom experience / Eugenia Bone
Description: First edition | New York : Flatiron Books, 2024 | Includes bibliographical references and index
Identifiers: LCCN 2024006622 | ISBN 9781250885654 (hardcover) | ISBN 9781250885661 (ebook)
Subjects: LCSH: Bone, Eugenia. | Mushrooms, Hallucinogenic. | Psilocybin. | Hallucinogenic drugs.
Classification: LCC QK604.2.H34 B66 2024 | DDC 362.29/4—dc23/eng/20240510
LC record available at https://lccn.loc.gov/2024006622

Our books may be purchased in bulk for promotional, educational, or business use. Please contact your local bookseller or the Macmillan Corporate and Premium Sales Department at 1-800-221-7945, extension 5442, or by email at MacmillanSpecialMarkets@macmillan.com.

First Edition: 2024

10 9 8 7 6 5 4 3 2 1

In Memoriam

Gary Lincoff (1942–2018) and
Tom Volk (1959–2022),
who taught three generations of
mycologists, including amateurs like me

Disclaimer

This book is for informational and educational purposes only and should not be construed as advocacy for the use of any illegal substance. The possession, sale, and/or use of psilocybin is illegal under United States federal law and the laws of many other countries. While decriminalization of psilocybin is underway in a number of state and local jurisdictions, the reader should consult legal counsel with regard to these matters. Moreover, the information in this book is not intended to replace the advice of the reader's own physician or other medical professional. You should consult a medical professional in matters relating to health, especially if you have existing medical conditions, and before starting, stopping, or changing the dose of any medication you are taking. Individual readers are solely responsible for their own health care and legal decisions. The author and the publisher do not accept responsibility for any adverse effects or consequences individuals may claim to experience, whether directly or indirectly, from the information contained in this book.

Contents

Prologue

Last summer, I was in my rural Colorado community's grocery store when a lady I know stopped me mid-aisle. She told me about her granddaughter, a high-achieving college student who was barely hanging on, her education upended by anxiety and depression. She wanted to know how to get her granddaughter the psychedelic drug psilocybin because she had heard a onetime dosage would resolve the girl's troubles. But she wasn't clear on the distinction between clinical trials and retreats, psilocybin versus magic mushrooms, and the efficacy of microdoses in relation to heavy trips. As the ice cream in my cart melted (Ben & Jerry's Cherry Garcia, the psychedelic flavor), I explained what I'd learned over the years: how to find a clinical trial, how to judge a retreat, the difference between doses. I also told her psilocybin, the active ingredient in magic mushrooms, wasn't a panacea and she (and her granddaughter) might want to manage her expectations. My neighbor thought for a minute and then, in her unembellished western way, said, "Well, that all makes good sense."

This is what happens when word gets out you are researching psychedelic drugs. I constantly meet folks who want help deciphering the hype. They seek plain insights that aren't tainted by advocacy or commercial interests or the pursuit of clicks, information that simply makes good sense. Never mind that just a few years ago in a small town like ours, trippers were seen as degenerates.

When I first got interested in mycology about twenty years ago and began attending mycological conferences and lectures, it was rare to hear anything about magic mushrooms and the psychoactive molecule, psilocybin, they contain. They were the black sheep of the mycology world. That's not surprising. When the Controlled Substances Act was passed in 1971, psilocybin—and by extension, species in the *Psilocybe* genus of mushrooms that contain it—was classified as a Schedule I drug, one with no accepted medical use and a high potential for abuse. Up until recently, it probably would have tanked a scientist's career to work on *Psilocybe* in any capacity besides a taxonomic project. People who were into these mushrooms were generally seen as druggies, or druggies pretending to be something else, like ethnomycologists, doctors, or spiritual leaders.

For decades, the only venue where magic mushrooms were discussed with any academic seriousness was the Telluride Mushroom Festival. For much of its history, the festival, which began in 1981, was small and funky. That's because there were so few people working on psychedelics in a serious way, but also because many saw the festival as an opportunity to trip. I remember a panel discussion where one panelist kept bursting into apropos-of-nothing giggles and another, teetering on his stool, pretended his mike was off and just mouthed his responses. The talk was, unsurprisingly, more hilarious than informative. But all that has changed. In 2023, the festival sold out. Notable scientists shared papers on their current research. Andrew Weil, one of the festival's grand pooh-bahs, who hadn't attended in years, offered a keynote address on how the psychedelic scene has evolved from fringe to mainstream. Speakers get swag bags now, loaded with mushroom products and psychedelic stickers, reflecting the boom of new mushroom-oriented businesses. And the audience, once scant and mostly inebriated, is full of eager listeners busily taking notes.

That psychedelics have gone mainstream was made clear to me at the 2023 Psychedelic Science conference in Denver. Widely and enthusiastically touted as the *largest psychedelic gathering in history*, it was more like the world's biggest pep rally for altered states.

The Psychedelic Science conference was sponsored by the Multidisciplinary Association for Psychedelic Studies (MAPS), a nonprofit founded in 1986 by the drug activist Rick Doblin that has become a leader in psychedelic research and education. (If, as anticipated, MDMA becomes available to treat veterans with PTSD in the next few years, we will have MAPS to thank.) Over the course of five days, thirteen thousand people wandered the chilly concourses of the Colorado Convention Center searching for the talk on Amplifying the Wisdom of Psychedelic Elders (room 201), Efficacy of Psilocybin in OCD (Four Seasons Ballroom), and Buddhism and Psychedelics (room 605). There were book signings, film screenings, and performances, all of them seemingly happening at the same time. I walked so many steps scuttling from presentation to presentation, my fitness app sent me a congratulatory message.

An exhibit hall housed a maze of some 275 exhibitors, hawking everything from flotation devices to psychedelic mushroom spore syringes for cultivation. All kinds of advocacy groups peddled their mandates, like PsillyGirls (harm reduction tools and artisan functional mushroom creations), Moms on Mushrooms (a support group for stressed mothers), and PooGod (exotic mushroom supplies). When it was time to leave the hall, I was so discombobulated, I had to be led out by a volunteer. A cavernous corner was dedicated to vivid art—much of it in Day-Glo colors—virtual experiences, and a raucous auction headed up by the multimedia artist Android Jones, a master of the psychedelic genre who looked like the kind of guy you would befriend in high school detention, only with a shaven head and a ponytail of skinny strawberry-blond locs, like a bouquet of limp #2 pencils.

Judging from the conference attendees, the demographics of the psychedelic scene are way larger and more diverse than the crowd I came to know all those years ago at the Telluride Mushroom Festival. Today, it includes bros who check the NASDAQ on their phones while gobbling MDMA, umpteen PhDs in something psychedelic, earnest fans who throng the scientists, BIPOC organizers and indigenous people who argue for equity and a "right relationship" regarding the use and distribution of psychedelic substances, and a large and curious public, many in sequins. Even Rick Perry, the conservative former governor of Texas, was there (though not in sequins) in support of psychedelic treatment for veterans suffering from PTSD.

At the conference, I noticed that many of the speakers started off by explaining their "context," sharing relevant personal background and socially important details like their preferred pronouns. I think this trend reflects the expectations of the greater psychedelic community, where individual experience is valued as much as academic credentials. And indeed, the significance of lived experience when it comes to psychedelic drugs cannot be overstated. Lived experiences have been shared for over a millennium among indigenous peoples, specifically those residing in Mexico and other Central American countries, and their knowledge about magic mushrooms remains the greatest reservoir of experience, followed by the collective experiences of the rest of us. Only a small percentage of what is known about these drugs has been generated by science. In fact, despite the headlines trumpeting breakthroughs, most psilocybin research isn't much further along than showing promise.

To understand how magic mushrooms are used today, it is essential to recognize the anecdotal experiences of magic mushroom users, the folks who consume the *Psilocybe* species that contain psilocybin and other psychoactive molecules. Their collective know-how represents an important body of knowledge from which we all can benefit—and not just regarding *Psilocybe* species. There are other species of mushrooms that are psychoactive, like *Amanita muscaria*, but while those users are less numerous, interest in Amma, as *A. muscaria* users call it, is growing, in part because it is a legal mushroom to consume. It is also a riskier mushroom in many ways, requiring pretreatments to modify the active ingredients, ibotenic acid and muscimol. That is an interesting story, but it belongs in another book.

Those who choose to use magic mushrooms do so for work enhancement, to treat mood and abuse disorders, to deal with trauma and stress, to enhance their spiritual lives, and to just have fun, either solo or in community. Many self-organize based on identity, like BIPOC groups, or in search of cures, like cluster headache sufferers. They find one another by word of mouth or through the internet. Some trip together, but many participate in a kind of after-the-trip group therapy, where they share their experience and try to make sense out of it with the help of other like-minded people. These citizen psychopharmacists are essentially doing research on themselves as indigenous peoples have for a very long time. And it is their reported experience that often leads the science. Research in microdosing mushrooms, for example, is happening because so many have said it has helped.

Have a Good Trip is a book informed by trippers and their trip reports (the practice of writing up the details of your trip and sharing them). Part 1 provides basic information about psilocybin-containing *Psilocybe* mushrooms and how they interact in your body, as well as what a bad trip is and how it can be mitigated. Part 2 details the different ways people acquire mushrooms. Finally, in part 3, I break down the three main trip types (microdosing, therapeutic, and spiritual), and then I look at the variety of psychedelic communities that are organized for support, pleasure, and sometimes both. The point of this book is to offer a basic understanding of what these mushrooms might and might not do.

This book is also a travelogue of sorts. In my attempt to understand how different people use these fungi, I used them, too, and in the process, I found out a lot about myself and how I interact with others. Not all of it was easy, but I can honestly say I am better off for it. Obviously, I can't give advice, but I can present what I've learned, agenda-free, just like I did with my friend in that Colorado grocery store last summer.

Author's Note

The psychedelic vernacular is developing at a faster pace than copyediting manuals can be revised. The language we use to describe psilocybin, the experiences it occasions, and the people involved in them lies in a murky realm between science, slang, and anthropology. Psilocybin, for example, may be referred to as a drug, a medicine, or a sacrament. Since the state of the vernacular today tends to be interpretive, I've made certain choices regarding nomenclature.

Many people object to calling psilocybin a drug, the word I use, and have taken to calling the psychedelic mushroom a medicine instead. *Medicine* has come to represent an antidote for everything from OCD to societal ills. But as the late psychonaut (someone who uses psychedelics to explore alternate realities) Kilindi Iyi said, medicine implies sickness. People who use mushrooms for mind expansion, spiritual quickening, or just to have fun are not using the mushrooms to heal something broken but to enhance their lives. Therefore, I use *medicine* only where it appears in quotations or applies to treating a physical or mental ailment. I understand the unease people feel with calling psilocybin a drug, but I don't want to characterize it as separate from other drugs that people use. Dr. Carl Hart, one of the world's foremost experts on the effects of recreational drugs, terms this differentiation *psychedelic exceptionalism*: the notion that psychedelics aren't like the dirtier, hard drugs. All drugs are psychoactive substances, said Hart on the *Bad Faith* podcast, "and we are all seeking the same thing when we use them. What we are talking about is making social distinctions with the drugs."

Lots of people use the phrase *psychedelic space* to describe anything from the personal boundaries surrounding a tripper to psychedelic-friendly advocacy and business categories, as in "I'm an entrepreneur in the psychedelic space." I use *psychedelic scene* instead because *scene*, to me, suggests an arc of activity, whereas *space* makes more sense to me when it is used to describe the protective envelope a trip sitter contains for the tripper. *Tripper* is another word that I use instead of *journeyer* or *voyager* or the like, and *trip* instead of *journey* or *voyage*. It's too bad that *tripper* has so much baggage, related, as it is, to the counterculture, but I think it just has more historical precedence. Another word I wrestled with was *ceremony*. Many people call their trips *ceremonies*. Except where someone is describing their intentions that way, I only use *ceremony* in specifically religious or anthropological contexts. I also tussled with how to describe the indigenous zones of mushroom use (basically, agricultural villages in Central America) versus the post-discovery zones of mushroom use (as in the US, Europe, and Australia, along with all their capitalist hubbub) and the different value systems they represent. I settled on using the *indigenous world* versus the *nonindigenous world*.

There is some disagreement about whether cultivars—the breeds of a particular species of mushroom—should be capitalized or not. I'm capping because reading *albino penis envy* in all lowercase just looks too weird to me. Mushroom genus and species names are always italicized—that's a well-established form. Also, as I describe in this book, when we metabolize psilocybin, it converts to another molecule, psilocin, and it is psilocin that causes the trip. For this reason, I wanted to use *psilocin* in the text, but I decided to bow to common practice and use *psilocybin* in most circumstances.

Even though it is a federally banned drug, some people are so enthusiastic about the benefits of psilocybin they are willing to take risks to promote it. One woman, a microdose consultant and passionate advocate of psilocybin, told me I could include her name, as she was "willing to go to jail for the medicine." But I said no. I masked everyone who asked and some who did not.

PART I

Anatomy of a Trip

1

Considering Psilocybin

Before I started this book, the last trip I had taken was in 2011. I took two grams of the most popular magic mushroom, *Psilocybe cubensis*, with a few other people in a meadow high in the San Juan Mountains. It was a beautiful Colorado day. The views were so long and so wide, I could see two rainstorms at once, though they were far away and many miles apart. The landscape rippled to the west, a giant wrinkled bedspread of red rocks and green pine forests. It would have been beautiful under any circumstances. On mushrooms, it was extraordinary.

My trip was mostly merry, a journey that took me from curiosity—*What is that scratchy noise?*—to amazement—*It's the busy lives of a million insects crawling through the grass, navigating the gargantuan obstacle of me*—to uncontrollable giggles as I realized I am as much a part of the landscape to them as the rocks are to me. Tripping like this, joyful and safe, defines the trips many people take today and the trips people have taken ever since magic mushrooms were first introduced to the non-indigenous world in the late 1950s. They were and are—or can be anyway—a way of altering one's perception that leads to a greater appreciation of nature, of the marvelous details we don't normally notice. How much more fascinating is a rain shower when you see it is composed of countless individual drops, glinting like little pear-shaped mirrors. How delightful to be alive amid all this!

There are and always will be meadow trippers, folks who use psychedelics to enhance their experience in nature, to free their minds from its bindings for a while. But just because someone is tripping in a meadow doesn't mean it's all sunshine and glistening raindrops. I've been in the woods with fellow trippers who spent the whole adventure puking or facing realizations about mistakes they've made with their families or friends and then wandered among their fellow trippers in search of someone coherent enough to help them talk it through. I've spent hours—or what seemed like hours—swatting mosquitoes off a comatose friend as he lay in the grass, traveling behind his sunglasses to foreign lands.

Some people have bad trips, too. Some even suffer psychotic breaks, though according to the scientific community, a person diagnosed with (or who has close family history of) schizophrenia or bipolar disorder is the most likely candidate. But we didn't know that at the hippie boarding school I attended in the 1970s. We didn't even have mental health counselors. We had a bad-tempered nurse who dispensed the narcotic Darvon for everything from menstrual cramps to learning your parents were getting divorced. There wasn't much talk about mental illness in those days. When a kid fell apart, we dorm-mates assumed they had become undone by psychedelics, which at that time were widely hyped as capable of frying your brain. I remember one girl who burst into rooms, long hair wild, pimples aflame, hyper and articulate and brilliant at math. One day she disappeared from school. Just gone. She wrote me a letter that said how much she missed her friends, and then I never heard from her again. We all thought she was sent away because she took psychedelics.

Today, I think she might have been diagnosed with ADHD. The art and science of mental health diagnosis has changed a lot since the '70s—some would argue, I think, that it's at times excessive. In the late 1990s, I met an administrator at a private grade school in New York City who encouraged parents to put their hyper little boys on ADHD medication, an amphetamine known around town as the "good-grade pill." A generation of teenagers have been prescribed antidepressants to deal with depression and then more prescriptions to deal with the side effects. A twenty-year-old relative of mine at one point was taking three different medications to treat her depression. She told me all those meds made her feel like she had lost track of her personality. It seems to me that the resurgence of mainstream study in psychedelics to treat mental disorders, beginning in 2006 after a thirty-year hiatus, has coincided with the evolution of mental health diagnosis and the subsequent need for new, effective mental health drugs. I know firsthand that the need is great.

In 2014, I wrote an op-ed in the *New York Times* arguing for the rescheduling of psilocybin, the psychoactive molecule present in a wide range of mushrooms but especially in the genus *Psilocybe*. Rescheduling psilocybin from its current status as a drug with no potential medical use doesn't imply blanket legalization. No one is talking about making psilocybin available at your local bodega or allowing children to purchase the stuff like gum, but it does have great therapeutic promise for an array of disorders and should be moved to a drug category that would make research less arduous and allow for compassionate-use provisions. Within hours of publication, I received hundreds of emails from people who were suffering from depression or PTSD or other disorders and wanted to know how they might get mushrooms. Even folks with complaints that didn't meet the standards of a diagnosis wanted to know more.

Psilocybin can positively impact nonsufferers of mental ailments. They did for me, and I wrote about it. The first time I ate magic mushrooms as an adult was in 2009, again on a mountaintop in the San Juans. I experienced several small epiphanies—the kinds of self-realizations that are the rewards of hours of therapy—but one in particular remained with me. As the drug wore off, I went back to my hotel room to take a hot bath. For a moment, I thought that might not be a good idea, as bath time is when women my age can be very self-critical, and I didn't want the sight of my thickening waistline to veer me into a bad trip. But while in the tub, I envisioned my body as a ship that was taking me through life, and that's what made it beautiful. With that change in perception, I stopped feeling regretful about losing my looks. Instead, I felt overwhelming gratitude for having a body at all. It was a tremendous relief that I still feel today. I received a letter from a gentleman after I published my editorial who wanted his wife to try the drug because she felt so badly about how age had ravaged her beauty.

The response to that op-ed was overwhelming, and I felt like I had irresponsibly unleashed hope where hope was truly premature. I tried to respond to the loads of correspondence from strangers who shared such intimate despair. I was deeply moved by their stories, but I really didn't have anything to offer. I did have coffee with one man—his ex-wife had reached out to me requesting I at least talk with him—and he cried the whole time.

I feel like I have so much more I could offer that fellow today. First, I'd make sure he understood there is no prescription. You cannot reliably eat magic mushrooms and expect whatever is plaguing you to disappear, like taking an aspirin for a headache. The scientists in this field are plugging away, trying to figure out what the abilities and limitations of the drug are, but that research is in its early stages. And the job is made all the more difficult because we do not all act the same under the influence of psilocybin.

But I'd also tell him there is a wealth of helpful anecdotal information if he is willing to accept the fact that different people experience the drug differently. Over the years, I have talked with hundreds of people who use these mushrooms for a variety of reasons, from entertainment to therapy to spiritual awakening, who have experienced personal insights into work and self, and have generously shared their experiences. I've also spent many, many hours reading the various online psychedelic forums, mostly Reddit and Shroomery. (*Forum* is singular, *fora* is the plural, but to avoid confusion I'll use *forums*.) The demographics of these forums seem to skew white and male, which limits the variety of experiences reported. Nonetheless, the subcategories (like r/badtrip) provide a wealth of views (and sometimes risky opinions) on all aspects of mushroom use and its effects that simply don't exist elsewhere. As the writer Ed Prideaux pointed out, "Go on to any psychedelic reddit page, and you're likely to find a much more genuine picture of what's happening in the 'psychedelic renaissance' than any asinine puff piece on CNN."

One thing is very clear: different doses make different trips. But regardless of dose size, the chances of positive outcomes are likely increased when trippers manage their expectations and employ caution. This can be accomplished by understanding the basic pharmacology (to the extent it is currently understood), the general effects that varying doses seem to occasion, the importance of a safe setting, and the companionship of trustworthy people. Psilocybin isn't mechanistic the way a statin is. It's neurological. For example, I benefited from the mushroom emotionally because the drug offered me the opportunity to see myself and how I interact with the world from a new perspective. That in turn affected how I felt about myself and others.

When I am writing a book, I usually have a novel that I read at night to relax. As is often the case, I encounter something that resonates with what I am thinking about during the day. So it was with the murder mystery *The Music of the Spheres*. In it, a character paraphrases Isaac Newton: "It is our greater or lesser knowledge of constant factors that constitutes their relative stability." In the book, he is applying it to astronomy. But it applies to tripping as well. What you can know and what you realize you can't know about psilocybin and its effects are a measure of the degree of confidence you can have in the experience overall. Ultimately, that's determined by Dose + Intention – Risk.

I am going to dive deep into these knowns and unknowns in subsequent chapters, but let's start with an overview of the key points when considering psilocybin: what the drug is and what it basically does, how dosing is measured to determine trip types, and the importance of setting an intention and making a plan to mitigate the risks.

The Basic Pharmacology

When we talk about magic mushrooms, we are usually talking about species of mushrooms in the *Psilocybe* genus, almost all of which contain the tryptamines psilocybin and psilocin. Tryptamine is a type of alkaloid, an organic compound that plants and fungi produce that has profound pharmacological effects on us. And indeed, these molecules are responsible for the psychoactive effects we experience. Other alkaloids are present in trace amounts, too, like baeocystin, norbaeocystin, and norpsilocin, though no one really knows whether they modify our trips. But I don't see the wisdom in dismissing them: I like to think of the magic in these mushrooms as a cocktail of psychoactive compounds dominated by psilocybin and psilocin. Psilocybin and psilocin are classic psychedelics, like mescaline, LSD, and DMT, and they all operate on the brain similarly.* Here's how.

After you swallow it, the stew of acids in your gut breaks the mushroom down. Acids break apart the psilocybin molecule, too. They strip off one of the molecule's components, the phosphate group. Psilocybin without its phosphate group is a different molecule, called *psilocin*. And psilocin is a molecule slippery enough to pass through the blood-brain barrier, that semiporous gateway between whatever is in your bloodstream and the ultrasensitive environs of the brain. So even though we say we are tripping on psilocybin, we're actually tripping on psilocin.

Psilocin is also present in the mushroom—though to a lesser degree. It passes through the blood-brain barrier along with the psilocin made from dephosphorylated psilocybin. The more psilocin naturally present in the mushroom, the faster the effects will come on because there is no conversion process happening in your gut—it goes right to the brain.

Psilocin is the molecule that causes the psychedelic effects

* Nonclassic psychedelics like ibogaine, ketamine, MDMA, and *Amanita muscaria*—the iconic red mushroom with the white dots—act on other parts of the brain.

we feel when we eat magic mushrooms, influencing our perceptions both sensory and emotional. That's accomplished because the chemical structure of psilocin is very similar to the chemical structure of the neurotransmitter serotonin, so similar that it can dock into serotonin receptors on neural cells and stimulate the transmission of electric impulses from cell to cell along serotonergic pathways in the brain. Your brain accepts psilocin as if it were serotonin, but of course it is not, and it doesn't do the same things. What it's doing in the brain, exactly, is hypothetical, but the prevailing notion is that psilocin affects—even disrupts—established patterns of electric impulses in regions involved with mood, memory, cognition, and the senses. It's as if, wrote the author Andy Letcher, "a new, alien but curiously compatible piece of software is thrown into the brain's computer, disrupting its normal operations in novel and unexpected ways."

A psilocybin trip typically lasts four to six hours, though an eight-hour trip is not uncommon. It doesn't cause a hangover like alcohol, but you may very well be tired or headachy after tripping. For me, after a trip, I feel mentally spent, like I've been reading all night, but emotionally opened, like a just-unclogged sink.

Different Doses Make Different Trips

A warning when it comes to different doses and the trips they may occasion. There is no one-size-fits-all dosing schedule. A big dose for me might barely register for someone else, and that's not necessarily because of gender, age (of adults), body weight, or metabolism. Those factors don't seem to matter, I suppose, since we all have the same size brain. Dose size cannot be predictably reproduced from one person to the next. My friend who took what is generally considered a heroic dose, five grams of a moderately strong species of *Psilocybe* mushroom, traveled in his mind to foreign lands. However, someone with a history of selective serotonin reuptake inhibitor (SSRI) use might spend the whole afternoon feeling frustrated, unable to let go, buzzy and distracted, but not tripping. No cool sensory experiences, no euphoria, no revelations: just disappointment in the drug and themselves for being unable to benefit from it. And someone who took two grams, expecting to feel one with nature without getting lost on the trail, might end up on their back for four hours in some kind of hyperspace of their own minds. All this could happen. All this has happened.

The only way to judge tolerance is to start with very low doses and, if appropriate, increase incrementally. This is called *titrating*, and it's one of the ways many trippers individualize their dose. That said, here are general dosing guidelines for *Psilocybe cubensis*. These guidelines are widely available on psychedelic sites and in books like *Medicinal Mushrooms* by Christopher Hobbs, an herbalist with the sober temperament of a high school health ed teacher.

A **microdose** is a subperceptual dose. That means you don't experience any psychedelic effects. It doesn't affect your daily routine. If you are microdosing properly, meaning you are consuming a subperceptual dose, then you can exercise and eat the way you normally do. You should be able to drive, attend to your job, go to the grocery store, help the kids with homework.

Microdoses are usually taken to manage mood or productivity. Those who microdose psilocybin to address mood claim it doesn't elevate their frame of mind so much as remove their negative state, that the minuscule amount of serotonin receptor stimulus they are getting brings them to a normal baseline, or to paraphrase Ayelet Waldman in her book about microdosing LSD, *A Really Good Day*, it just helps you feel not shitty. It also helps some people cope with the stresses of normal life and stay chill. Those who microdose magic mushrooms for productivity claim it helps them focus and stay energized at work. I found that to be true for me.

A microdose applies to any psychedelic, but in the case of magic mushrooms, it is considered 50–500 milligrams of dried *Psilocybe cubensis*, the most used species, ground fine to measure. For me, a dose of over 160 milligrams is perceptible, depending on the cultivar of *P. cubensis*. (More on that in a minute.) There are many scientists who are skeptical that microdosing does anything at all and believe the benefits people claim are due to the placebo effect. That may be true. But a plethora of anecdotal testimony suggests microdosing does have an effect, so much so that researchers haven't given up on it. On the other hand, lots of people swear by homeopathy, too, and there's no evidence it does anything.

A **low dose** between 500 milligrams and 1 gram may produce euphoric feelings and sensory enhancement. It can make you feel a little high, a little more enthusiastic, a little more excited, a little more appreciative. A friend of mine, a real estate guy, took about half a gram of a strong *P. cubensis* cultivar and went to the ballet—usually a bit of a chore for him, but his wife likes it—and he said it was the best, most incredible time he'd ever had at the ballet. One to 2.5 grams may produce some psychedelic effects as well. In mid-twentieth-century clinical trials of LSD, frequent low-dose use called *psycholytic therapy* was studied for its therapeutic value. These are, for many people, manageable doses—what Christopher Hobbs described to me as a **museum dose**, meaning you can go to a museum and enjoy yourself without freaking out. Some use low doses as a substitute for drinking on a night out—you get high, but there's no hangover. But doses this size can be tricky. Many times, I've heard stories about people who have taken 2 grams of magic mushrooms, thinking they'd just have a super time at their wedding reception, a music festival, an afternoon at the beach, and it's turned out to be a full-on unable-to-manage-things-like-making-change-at-a-store kind of experience. I think that's because at a dose this size—and this is true of microdosing as well—the psilocybin content of different cultivars can make a substantial difference.

Of all the *Psilocybe* species that contain psilocybin and psilocin, *P. cubensis* is the most used because it is easily cultivated. But even then, there can be differing potencies of *P. cubensis* cultivars. Albino Penis Envy, for example, is significantly stronger than Golden Teacher. It's like horse breeds. A Clydesdale and a Shetland pony are both *Equus ferus caballus*, but one is four times as heavy as the other. Someone taking 2 grams of *P. cubensis* may think it's just a museum dose, but if the cultivar is Albino Penis Envy, the amount of psilocybin that 2 grams delivers may be significantly more. At this intermediate dose level of 2 grams, I've had a significant enough experience where I didn't want to move, but I've also had a very light experience, where I could go to a café and order a coffee. Maybe it had something to do with the cultivars, but on the other hand, maybe it had something to do with that ultimate wild card, my frame of mind at the time. In the stronger experience, I had fasted and participated in a subdued and ceremonious nighttime event. In the lighter experience—or maybe this made the experience lighter out of necessity—I wanted to keep an eye on my husband, who was on a deeper trip. One characteristic of low-dose tripping is that sometimes you can pull it together if needed. On a high dose, that may be more difficult.

A **high dose** is generally considered to be 2.5–5 grams and will likely produce psychedelic effects like visual and audio distortions when eyes are open, and vivid waking dreams, memory retrieval, and stream-of-consciousness narratives when eyes are closed. It can cause more intense euphoria and possibly ego dissolution, a phenomenon associated with mystical experiences and mental breakthroughs. Two and a half or 3 grams dried *P. cubensis* mushroom—about the equivalent of 25–30 milligrams synthetic psilocybin—is a typical **therapeutic dose** used in clinical trials, retreats, and ceremonies. In conjunction with therapy, a high dose of psilocybin has shown positive effects for some people suffering from illnesses like addictions, treatment-resistant depression, OCD, and other disorders. That's what the rapturous headlines you have been reading refer to. People who take high-dose psilocybin often speak of the benefits of integrating their trip by talking with someone afterward, either a peer group or a trained therapist, and examining their experience to articulate whatever learnings they believe can be gleaned. It has been characterized as an "unfolding process . . . a continuous unraveling of insights about oneself and one's relationships which can take place over weeks or months." This can be especially helpful if you are distressed after a bad trip.

On a dose this size, it is common to feel some anxiety coming up, and it may linger, even take your trip sideways. Likewise, magic mushrooms, really on any dose, can bring up a lot of emotional stuff. I know someone who took a microdose of 150 milligrams and, while driving to the hardware store, became overwhelmed with nostalgia and burst into tears. He had to pull over to recover. Emotional bloodletting of this sort isn't uncommon, and for some people, it is cathartic, but of course it can also be confusing and feel inappropriate. On a high-dose trip, it's a good idea to have a trusted trip sitter around, someone who can remind you everything is okay, that the trip is temporary, and generally make sure you don't do something stupid.

Five grams and up is known colloquially as a **heroic dose**. The term was coined by the ethnobotanist Terence McKenna. A heroic dose can lead to ego dissolution, alternate realities, and challenging but transformative insights—not the kind of trip to do at a bar mitzvah. On a dose of this size, you may forget you are tripping and transcend your existence altogether. But these strong trips can be tough. It's not uncommon to throw up or lose bladder control. One tripper I spoke with said he uses magic mushrooms "periodically, for a burning bush moment." But once he did 15 grams of *P. cubensis*. "That was bad. I got really scared, my temperature spiked, I had to get into an ice shower, I puked. For me, no astral projections. It was like having the flu." People have described terrifying experiences, too. In one trip report I read, a fellow visualized hanging himself. Another spent his trip watching the world blow up over and over again. But people have also described ecstatic revelations about how to live their lives and insights into the nature of, well, everything. Such life-changing realizations can even lead to course redirection, which is why you should never do something dramatic after a trip like telling your spouse you want a divorce. It makes sense to me to sit on a revelation for a while before hitting Send.

But don't confuse a heroic dose taken as a "tool for exploration" with a double-digit dose taken because you can't get off on a smaller one, as occurs with some people. People who take heroic doses usually expect a big light show. To those who must consume lots of psilocybin to feel any effect at all, there's nothing heroic about it.

By the same token, it's not so unusual for people to brag about how much psilocybin they took regardless of whether it was planned or not. It strikes me as a kind of chest thumping, like competing in a hot chili pepper–eating contest. The last thing anyone should do, in my opinion, is judge themselves based on the amount of psilocybin they've taken. More than once, I've had trippers brush me off when I've tried to express my psychedelic experience on 2 grams, like it really wasn't an experience at all, and I was just embellishing. At the time, it made me feel inadequate. When I asked Dennis McKenna about the implications of large versus small doses—he's kind of the King Solomon of psychedelics, an ethnopharmacologist specializing in hallucinogenic plants and brother of Terence—he said it takes courage to confront your inner fears and trepidations at any level of consumption. So that's what I keep in mind: how *you* perceive your experience is the only thing that matters, because the effects of a given dose are ultimately determined by each person's own brain chemistry.

Setting Intentions

When I was experimenting with psychedelics in high school, we didn't think about what kind of trip we wanted or what a particular dose would do. Our intention was to see what would happen and hopefully have a good time. I have a friend who had his first trip at eighty years old. He decided not to set an intention at all, because he didn't want to hem in his experience in any way. (By the way, it turns out that as you age, psychedelic trips can become mellower.) Lots of people trip for mind expansion or just to have a good time. I've been out hunting psychedelic mushrooms with other mushroom hunters and seen them pop raw mushrooms in their mouths as they walked along, just like you do when strawberry picking. There was no intention other than to find mushrooms and have fun.

Those who are interested in using psychedelics for specific purposes, however, understand the importance of intention. Setting intentions, having a goal for the trip, "may be the single most important factor in determining individual reactions," wrote Andrew Weil. It's important because when you determine why you are taking the drug, it not only defines what kind of dose to take but what kind of results you are after. That's a prescriptive approach to the drug, and it's what has so many people interested in it today.

People often use intentions to seek transformation. Goals such as *I want to quit drinking, How do I fix my marriage?* or *What path should I follow in my life?* can anchor a trip. Some folks like to describe intention setting as asking the mushrooms, and the mushrooms give them the answers they seek. That's something like the indigenous approach, where the mushrooms are associated with divine wisdom. I see intention setting as opening doorways in the mind that may allow you to see answers and solutions that have always been there but for whatever reason haven't been obvious before.

It's not uncommon for trippers to interpret images or scenes they experience as a metaphor for their intention. Those metaphors can turn out to be helpful tools going forward. Brian Richards, a scientist working on psychedelic therapy for cancer patients, told me a story about a patient in a clinical trial who had a successful (doctors say *optimal*) treatment for breast cancer, yet she was plagued by hypochondriacal fear. "Every signal of pain and discomfort occasioned panic as possible fear of reoccurrence," he told me. During her trip, she set an intention to go hunting for her fear, to find out where it was hiding in her mind and what was behind it. While tripping, "she opened a door and inside was a little cowering hedgehog. *This* was her fear? She held it and realized it was just trying to keep her safe from harm, to prepare her as best it could for any future suffering. From then on," he said, "she had a personally meaningful symbol to relate to that helped her discern and attenuate her symptoms."

Setting intentions is an established step in many clinical trials, and it seems to be derived from a far older practice. We have only been talking about magic mushrooms in this country for seventy years, but long before they came to the attention of the nonindigenous world, they were used in ceremonies by the Mazatec people of Mexico and native people in Central American countries. Villagers with a problem—perhaps a physical or mental complaint—may attend a magic mushroom ceremony with a *curandera*, or shaman, in the hopes that under her guidance and the spirits of the mushrooms they might find some relief.* They are not going to the *curandera* to get high, nor do they have a wealth of outlets for modern health care. They have a specific intention, which the *velada*, or ceremony, is designed to address.

If a tripper goes on a psychedelic retreat or hires a therapist or guide to work with them—both in preparation for a trip and

* Diagnosis via mushrooms is not as far out as it sounds. On the Reddit page r/shrooms, there are numerous testimonies by folks who, during a trip, identified a problem in their bodies that upon visiting their GPs turned out to be basal cell carcinoma, arrythmia, thyroid cancer, and more.

in therapeutic conversation afterward—the chances are he will be asked to set an intention. I took a variety of trips researching this book, and throughout, I was asked to consider intentions, to come up with questions "for the mushroom." I've already spoken of how psilocybin helped me get over losing my youthful looks, but I didn't ask for that insight; it just came at a time when it mattered to me.

Throughout my experiences, I had revelations about people I love and what they need from me and warnings about things I do that I shouldn't. But when it came to a specific intention, I didn't know what I wanted or needed. I knew what wasn't working for me in my life: *I'm dealing with getting older, yet I have no idea of how I fit into the scheme of oldness.* At my age, my grandfather had fully embraced Sansabelt pants and the golf cart life. I may be old, but not 1975 old. Still, I was on the verge of being a senior citizen, and it felt like I was aging out of contemporary culture. Becoming the human equivalent of VHS made me uptight and antisocial. Yet somehow examining my dread about the normal process of aging seemed unworthy, a waste of a good trip. But it turns out that nothing was wasted. Over the course of a year, tripping in various ways, I experienced a subtle but profound change in my attitude about everything, including aging. So even if you don't have an intention beyond generally feeling like you are not comfortable in your skin, that may still turn out to be intention enough.

Mitigating Risk

Psilocybin is widely reputed to cause no harm to your physical body. That might not be 100 percent true. Serotonergic agonists—drugs that stimulate serotonergic receptors—do have the potential to induce valvular heart disease, and scientists will likely have to study this to determine what constitutes safe dosing, especially in cases of frequent use, as in microdosing. An individual may be at risk based on the state of his or her mental health in general—people diagnosed with schizophrenia or bipolar disorder, for example, are discouraged from using this drug—and even one's state of mind in the moment. For example, tripping while sleep-deprived is probably a bad idea. But the most well-known risk is of self-harm occurring during a trip, or harm that arises from bad judgment, like driving your car while hallucinating.

Doing something potentially harmful to oneself or others may be mitigated by ensuring there are sober, competent people present who will interrupt any behavior that might be dangerous, and in the process confer a sense of safety. Most trips start with a twinge of anxiety. Those anxious feelings can dissipate if the setting feels secure, or they can accelerate if there is something or someone around that feels creepy. Anxious feelings may even precipitate a bad trip, not unlike a full-blown anxiety attack. Who a tripper surrounds him- or herself with can be a tremendous asset if they help assuage scary feelings as they come up, but they can also be the problem if their intentions for being there are prejudiced by their own self-interest. (Of course, one can have a tough go of a trip regardless of how supportive the sitters are.)

In a clinical trial, one's confidence level in the sitters may be higher, because in most cases, you will be in the hands of professionals with licenses to protect. In a retreat setting, it's iffier. Most guides I have encountered are mindful, well-intentioned folks, but there is always the possibility that a guide won't know they shouldn't impose their cosmology or worldview on you. It might not be helpful for a guide to tell you during a trip that the reason you are experiencing so much anguish is that you were a Nazi in your past life. Likewise, a trip can go south if a guide has ulterior motives for being with someone in that vulnerable state (watch out for horny guides) or some kind of agenda to fulfill (watch out for evangelism). If I have a bad gut feeling about someone, I am not going to give myself over to their judgment.

Equally important is the setting. One of the first warnings I heard about psychedelics was "trip in nature rather than the city." Magic mushrooms can enhance your experience of the outdoors, but you can also get lost or overextended physically. My husband told me a story about how, as a young man, he tripped on a campout and somehow lost his pants; not the worst thing that can happen, but think about it: How do you lose your pants? Lots of happy trips in nature involve staying put in nature. I remember as a teenager tripping in a blooming apple orchard. I felt safe and was preoccupied for hours by the soft pink flowers, periodically tossed in the air by a warm breeze, the fat bees buzzing about in slow motion. Music festivals and raves can create a sense of safety in numbers and like-minded users, but I've also heard many stories of bad trips that occurred when folks lost their friends in the crowd. And cities are chaotic places, full of variables and distractions and police officers who might not recognize your odd behavior is the result of a psychedelic, not psychosis. The two can look similar. A good setting—or the right setting for you—can make a world of difference.

The setting of a professional clinical trial may provide a sense of a safe space. In most cases, you can have confidence there is quick access to medical personnel if needed (usually not). But not many people will get into a clinical trial; if they can afford it, they will go on a retreat. Many retreats are held in beautiful vacation settings, but ultimately how secure those settings are is relative to the number of guides present to oversee all the trippers. If a place seems skeevy to you, I'd recommend eating the money you spent and just going home. Alternatively, many folks make private arrangements to trip at home, or in the home of a friend, settings that are animus-free. In the past, I've worried about things like, *What if the FedEx guy shows up? What if my landlord calls?* For such eventualities, it's a good idea to have someone around to run interference. We've all forgotten how to turn off our phones, that we don't have to respond to everything instantly. A trip is a good time to remember.

What you can't forget is that in most places, psilocybin is illegal. The potential of psilocybin to treat mental disorders has led to a general social relaxation about it, to the point where it *seems* to be legal, but that attitude can backfire in the wrong time and place. If psychedelics are something you think more people should have safe access to, then write your state representatives. You won't be the first person to tell them so.

A trip is determined by Dose + Intention – Risk. Understanding the relationship between the species of mushroom, the size of the dose, and your intention in taking them can increase the chances you will make good choices. You might not want to bother with microdosing if you are searching for a mystical experience: you may just end up organizing your sock drawer. You probably don't want to take a heroic-size trip if you are hoping to be more effective at a brainstorming meeting, as you could end up lying under your desk with your jacket over your face. Similarly, understanding the known risks, and seeking to mitigate them, also increases your ability to make good choices. It's too bad the science is not ready for the many people who have expressed an interest in psilocybin, and unfortunately, anecdotal knowledge doesn't rise to the same reproducible standards as scientific inquiry. But there is a lot of human experience using magic mushrooms, and much of it may be useful, if you know how to evaluate what trippers say.

2

Brains on Mushrooms

I have a history of ruminative thought patterns, especially at night. When it comes to reliving my public screwups, my memory is decades-long. Rather than sleep, I tended to list them, like counting sheep except that instead of inducing sleep, it prevented it.

Ruminative ruts like mine may be alleviated by psilocybin, and I think the handful of trips I took over the course of a year may have helped me. I'm not staying up all night anymore, berating myself over flubs from long ago. But how did that happen? How does psilocybin, once converted to psilocin, alter our thought patterns? How does it help people suffering from a range of disorders like OCD and PTSD? Why do we experience time warps while tripping? Yes, I have a lot of questions.

If only there were definitive answers. Despite our expectations, science is not about the truth so much as about searching for the truth. Today's proposed explanation for how psilocybin works in the brain can change tomorrow as new research emerges. I like to think of science as a verb: it's inquiry, always ongoing. That's frustrating for someone who likes a pat answer, but once I embraced the fluidity of scientific investigation, I was better able to read the current research without frustration. I did have to learn some basic biology, which I'll share with you, like how electrical stimuli move among different regions of the brain, but it turned out to be worthwhile because, well, learning always is. With that in mind, let's break down the anatomy of a trip as it is understood today, from the action of the molecule in your brain, to the effects it causes, to the therapeutic benefits it may occasion.

The Pharmacology of Psilocybin and Psilocin

I've already described what happens immediately after you swallow the mushrooms: the masticated bits encounter the stew of acids in your guts that metabolize the psilocybin into psilocin, and the psilocin travels in your bloodstream to your brain, where it interacts with your neurons.

Let's get a bit more granular. Neurons are nerve cells, the fundamental unit of our brains and nervous systems. Neurons use electrical and chemical signals to relay information between different areas of the brain, and between the brain and the body. The basic neuron is shaped like a many-armed starfish with one especially long arm. The shorter arms are called *dendrites*, and they receive messages *to* the cell. The long arm is called the *axon*, and it transmits messages *from* the cell.

Neurons are not physically connected. There is a microscopic gap—called the *synaptic cleft*—between the message sender (the axon) and the message receiver (the dendrite). Like little mailmen, neurotransmitters carry the message across the gap between cells. Neurotransmitters are made in the cell and captured in saclike structures called *vesicles*. When an electrical signal travels down the axon and reaches the axon's terminal, it triggers the vesicle to fuse with the cell wall and then discharge its payload of neurotransmitters into the synaptic cleft. The neurotransmitters flood the cleft, dock into receptors on the receiving cell's dendrite, and reestablish the electrical signal so that it may continue along its path. The juncture where brain cells interact and information is passed is called a *synapse*. It's where one neuron hands the baton to the next, like in a relay race.

Neurotransmission is not an unbroken chain of signaling, however. It's not like once a pathway from neuron to neuron gets started, it's off to the races. Synapses can make the cell more active or less active; for example, at a synapse, the electrical signal is stopped and assessed: it may get a green light to continue to the next neuron or a red light to stop. That assessment happens at every cell along the way. It's one of the factors that make brain communication so complex. Psilocin may influence that assessment when it docks into serotonin receptors, leading to pathways that don't normally occur in your brain.

As far as your brain is concerned, psilocin *is* serotonin. Psilocin travels passively in the bloodstream, slips across the blood-brain barrier, and then docks into various serotonin-friendly receptors. Serotonin is a neurotransmitter that is made by only about one million of the eighty billion neurons in the brain, but those one million cells send serotonin all over the brain and modulate a lot of behavior, pointed out the neurologist Jonathan Rosenthal (over the years, he's kindly explained—and reexplained—brain basics to me). Though serotonin has a range of functions, regulating things from mood to the movement of waste through your body, the serotonin pathways that are most affected by psilocin are those governed by particular receptors.

Our neural cells have numerous receptors, but psilocin is thought to interact primarily with two serotonin receptor types, receptors 5-HT1A and 5-HT2A, both expressed in the prefrontal cortex of the brain. Two giants in psychedelic science, Robin

Carhart-Harris and David Nutt, proposed that the 5-HT1A signaled pathway promotes passive coping; it keeps us mellow and able to manage stress. The 5-HT2A signaled pathway, which psilocin has a particular affinity for, is responsible for the psychedelic effects. This receptor's signaled pathway is characterized by cognitive flexibility and neuroplasticity (the ability to learn or unlearn), spirituality, and empathy.

That doesn't mean psilocin doesn't act on other receptors, though; it does. Psilocin has a low affinity for dopamine receptors, for example, which affect movement, emotions, and the mesolimbic, or reward, system. But so far, the 5-HT2A serotonin receptor is the most studied in relation to psychedelic effects. It's still to be determined how or if the interactions of psilocin with other receptors impact the psychedelic experience. The other psychoactive molecules in the mushroom may interact with serotonin receptors, too. Baeocystin, for example, dephosphorylates in the body (like psilocybin), hypothetically becoming norpsilocin. Researchers have found norpsilocin does indeed interact with 5-HT2A receptors, but it degrades faster than psilocin and maybe isn't that big a deal. Nonetheless, it may lend something to the effects of magic mushrooms.

The study of what psilocin does in the brain is constantly evolving. Just a few months after I finished this chapter, a paper was published in *Science* that suggests maybe psilocin doesn't act only on serotonin 5-HT2A receptors on the receiving cell membrane but on 5-HT2A receptors *inside* the cell as well. The same quality that allows psilocin to cross the blood-brain barrier—it's greasy—may be how it is able to slip into neural cells, too. Serotonin may not be the natural ligand, or partner, for 5-HT2A receptors inside the cell.

But psilocin, once inside that cell, does seem to engage

those receptors and stimulate neural growth. This potentially represents a new drug design for the treatment of depression. Neural growth occurs when neurons extend their dendrites and axons, allowing the cell to find other neurons and form more complex networks, which is indicated in a healthy brain. Fewer neural connections are indicated in depression. So, a drug that stimulates neural growth could lead to healthier brains. Medications that address serotonin receptors on the outside of the cell, like SSRIs, might be missing the target: serotonin receptors inside the cell. If you follow this stuff, it is major news indeed. Fascinating, too, because it leads one to wonder, if serotonin doesn't excite intracellular receptors, what neurotransmitter (besides psilocin) does?

The excitation of these receptors is just the beginning of a brain-wide web of hierarchical processes. From the point of excitation of the 5-HT2A receptor type, neurons connect to neurons, which cluster into distinct regions that perform various tasks. A frequently cited paper published in 2011 by Thomas Yeo and colleagues theorized that regions of the brain—and there are many—coordinate into networks that work together to perform a complex function. Take the visual network. It is composed of regions that have separate jobs involved in visual processing. Combined, those regions create a network that allows us to see the world the way we do. The 5-HT2A receptors that psilocin fits into so neatly are dense on neural cells that function in regions of the brain associated with adaptability, learning, cognition, and perception. It is the activation of those receptors, their neurons, and various regions and networks that is responsible for the effects we feel when tripping.

Identifying the regions of the brain that are excited by psilocin is an ongoing effort. The Centre for Psychedelic Research at Imperial College London has produced several proposals regarding which regions of the brain are lit (or unlit) by psilocybin, using functional magnetic resonance imaging technology. fMRI measures oxygenated blood flow in the brain. It's an indirect measure of brain activity in particular regions and networks. The scientists take pictures of the brain on psilocin and off, compare them, and interpret the images much as a radiologist interprets an MRI of your blown-out knee. These studies have shown that on psilocin, certain regions of the brain experience a reduction in activity, while other regions of the brain experience an increase. Some regions that normally communicate with each other decline in communication, and others that usually don't communicate much at all start communicating. Psilocin seems to be responsible for those changes. And all that activity can cross both lobes of the brain, leading to novel global integrations of mental processing.

Which regions of the brain are lit up by psilocin and which are downregulated has not been fully established. One consistently implicated area of the brain under the influence of psilocin, however, is the so-called default-mode network (DMN), described as a rambling network composed of many regions that is loaded with 5-HT2A receptors. The DMN is thought to play a role in a bunch of abstract functions like introspection, mind wandering, autobiographical memories, executive functioning, multisensory integrations, and decision-making. That includes a node within the DMN, the posterior cingulate cortex, thought to be associated with self-referential mental processes, "how we relate to our own thoughts and feelings."

Under the influence of psilocin, fMRI measurements of the DMN show less connectivity among the regions of the brain that comprise the DMN, but *increased* integration among regions *after* the drug has worn off.* Some folks have reported a reduction in their depression symptoms, and researchers think that might be due to the post-trip integration of neural pathways within the DMN. So what's going on here? Dr. Carhart-Harris has described the brain as a snow globe: the trip shakes up the snow, creating a sort of chaos, but when the snow settles, it lands differently, settling into new patterns. Hypothetically, when the DMN is downregulated by the drug, so, too, are ruminative thoughts or excessive self-focus—behaviors that are like wheels turning in a rut. Anybody can experience this, but these symptoms are indicated in depression, addiction, and anxiety disorders. Hypothetically, when the drug clears our brains, the DMN upregulates without symptoms like ruminative thought patterns . . . for a while anyway. When you hear folks say psilocybin "rewires" the brain or that it "rebalances" or "resets" the brain, they are talking about this refreshed integration of brain regions within the DMN. If you prefer the computer metaphor, it's like a reboot, which clears the cache and allows the program to run smoother upon restart.

Increased connectivity between some networks (other than the DMN) may happen as well, like the networks involved in sight and sound functions. But maybe the opposite is true and basic sensory networks become *less* active and specific areas of high-level networks like the DMN become *more* active. Depending on the study methods used, there are findings that

---

* It may be that psilocybin downregulates another part of the brain, the claustrum, a thin sheet of gray matter buried inside each hemisphere that connects to almost every region in the brain. The claustrum functions, suggests the neuroscientist Fred Barrett, as a "switchboard," capable of routing, rerouting, or not routing neural pathways in the brain. The claustrum is dense with 5-HT2A receptors, and Barrett and his colleagues at Johns Hopkins think psilocin might disorganize the claustrum, decreasing its activity, which in turn quiets the DMN.

support both observations. The bottom line is, since serotonin receptors are so widely distributed among different brain regions, the amount of cerebral real estate (and the functions it represents) that can be affected by psilocin one way or another is huge.

That summarizes the basic neuroscience as it is understood today. So far, there is no definitive answer about what is going on in the brain on psilocin. Someday, however, scientists may be able to point to a specific activation or deactivation of a particular region of the brain or specific connections made or unmade between different regions to determine which are responsible for the specific effects we feel when tripping.

The Psilocybin Experience

In general, you will likely start feeling the effects twenty to forty minutes after ingestion. The effects will intensify over the next sixty to ninety minutes, peak for a few hours, then wind down for another hour or two. As I've mentioned, a magic mushroom trip typically lasts four to six hours, though eight hours is not unusual, and the drug is evacuated from your system in about fifteen hours.

According to Stanislav Grof, a psychiatrist and a seminal researcher in the use of psychedelics for psychological healing, a tripper's consciousness consists of a sensory stage, where you experience hallucinations; going deeper, a psychodynamic stage composed of memories or past experiences (sometimes traumatic ones); deeper still, a perinatal stage that includes death and rebirth; and deepest, a transpersonal stage, when the tripper, no longer occupied with herself, acquires a sense that there is a *beyond*. You won't necessarily experience all or any of these stages, or in that order, but my observation is that the higher the dose, the deeper you plunge into your consciousness. However, the higher the dose, the likelier you will experience some kind of anxiety, too.

Psychedelic trips are entirely subjective and dose-dependent, so it's impossible to say definitively that X will happen and Y won't. But I can share with you the range of effects people have reported experiencing from onset to peaking to coming down and, where it applies, what it was like for me.

Onset

During the onset of a trip, that first half hour or so, it's common to feel an accelerated heartbeat, skin flushing, the shakes and shivers, and to experience yawning, giggles, headaches, or nausea. Lots of people vomit. For me, it's always the same: my heart starts thumping and I get the chills, which are like the postpartum shakes I experienced after childbirth. I also feel a little (some people feel very) anxious, a queasy feeling somewhere between mild nausea and worry, where I question whether this is something I really want to do. I find this is a good time to remind myself a trip is temporary.

Peaking

Once the experience is in full swing, a lot of different kinds of things can happen.

Your pupils dilate—as big as black olives—which can make

your eyes sensitive to light. Aches and pains may disappear, only to come back as the trip concludes. Prior to one trip, I had been suffering from neck pain for weeks. It was a relief when the pain melted away as my trip intensified. When the pain returned, I knew I was heading back to reality.

Trippers may laugh and they may sob. I once shared a seat on the bus to the Burning Man festival in Nevada with a bouncy gal from Southern California. I asked her if she intended to do mushrooms. No way, she said. "Last year, I took mushrooms hoping to have a euphoric experience on the playa [the alkaline lake bed where the festival takes place] and ended up crying in my tent the whole time. This year, I'm taking MDMA."

Some trippers may behave in irrational ways that can be alarming: stripping, threatening themselves or others, peeing on the floor, or becoming despondent. One person who posted on r/shrooms wrote himself a bad trip note, in anticipation, I think, of just these kinds of symptoms. In the note he reminded himself he was tripping, that the trip would pass, and to start thinking positively. "Since I myself wrote it . . . ," he noted in a comment, "it helps me believe I'll be ok since I always trust in sober me."

Visual distortions are typical. Technically, a hallucination is defined as perceptions arising from the absence of external reality. Extreme open-eye hallucinations, like mistaking a plastic bag for your cat, are less likely with psychedelic mushrooms but can occur, especially on very high doses. But what mostly happens when you are tripping with eyes open is sensory distortions of what is already there. Trippers may see rainbows and spectrums, and colors saturated and radiating with energy. You may see the world through fractals, lattices, honeycombs, tunnels, funnels, and spirals. Patterns may emerge from staring at a grassy lawn, a pebbly road, sand. I've watched clouds develop symmetrical patterns, constantly changing, like a kaleidoscope in shades of white and pink and gray. Order emerged from the chaos of the cloud forms, manifesting a kind of natural geometry, like seashells and snowflakes, and I felt an awesome sense that a universal order underlay *everything* in nature. You might find yourself attracted to and curious about all manner of things in your environment, and you may see them in a different light, affording insights into the complexity of the natural world. On psilocybin, things I don't normally notice, like the solar tracking of daisies, become seeable, a staggering reminder of how alive they are.

On higher doses, significantly stronger effects may occur. You may experience visual distortions like liquefying walls or items changing size. With closed eyes or in the dark, you may find yourself watching memories from your past, or scenarios may play out like in a dream. One tripper told me about encounters with shimmering animals in a forest who spoke with him. Another described being inside a spiral architecture, "like the Guggenheim Museum in New York," where on the walls, films played scenes from her life that she could stop and watch. I've heard stories of trippers seeing masses of threads or ropes, each representing a storyline in their lives, or a story they tell themselves about their lives, that they could pull to reveal. These stories may include conversations with "entities," sometimes wise and beneficent, but sometimes harsh, malevolent, or intimidating. Many people will speak of conversations or interactions with elfish beings that they interpret as the mushroom itself, what the Mazatec *sabia*, or wisewoman, María Sabina called the "holy children" or "the little saints."

Trippers may have a heightened sense of smell, taste, or

touch. They may have auditory distortions and enhancements of sounds already present. One person on PsychonautWiki described hearing their hair when they ran their fingers through it. In a group trip at night, I couldn't see anything, but I could hear other people in the large, dark room swallowing, rustling, breathing, sniffing, murmuring, all the sounds drawn out and modified into a low *wah-wah*. I couldn't tell if they were near or far; I felt surrounded by a thick, muffled wall of random human sounds. But I found I could focus on one voice and hear that voice intimately: a woman reciting prayers to herself in Spanish, another who was having trouble breathing through her nose. It felt a little like spying.

Synesthesia and Time Warps

Another typical tripping experience is synesthesia. Synesthesia is the phenomenon of experiencing one sense through another, where you might hear shapes, see tastes, taste colors, feel smells. Synesthesia isn't exclusively an effect of psychedelics: an example of non-drug-induced synesthesia is grapheme-color synesthesia, where letters or numbers or even musical notes are perceived as colored. In the tiny percentage of people who experience non-drug-induced synesthesia, the effects are quite consistent. For them, if the number 4 appears red, it will always look that way. The phenomenon is not really understood but is thought to arise from cross-wiring between processing segments of the brain that are not normally networked—just what seems to be happening when you are tripping.

Some scientists have proposed that all of us may start out life synesthetic, but as we mature, the brain settles into fewer, more efficient pathways. And indeed, fMRIs of the brains of some people diagnosed with autism, the brains of people tripping on psilocybin, and the brains of young children all show higher activity between regions of the brain than the neurotypical, the adult, and the sober.* Neural connections in the typical adult mind seem to be more streamlined: there are fewer connections between regions and within regions of the brain, but they are strong and fast. Think of adult neural pathways as a network of superhighways linking regions of the brain, in contrast to the meandering footpaths of a child's neural connections. That efficiency is beneficial in many ways; it saves energy and promotes focus. It also affects how we perceive time.

It's typical for one's sense of time to be warped while tripping. A trip often seems way longer than the six hours or so it takes for the drug to wear off. A short walk can seem to take forever, but then, when you aren't tripping anymore, it will turn out to be no distance at all. That's happened to me many times in the mountains. I will walk and walk and walk in search of the perfect spot to have my experience, but eventually get so tired and overwhelmed, I just plop down on the edge of the trail. After the effects subside and it's time to head home, I always psych myself up for what I expect to be a long hike. Inevitably, it's a matter of yards.

Marc Wittmann, a neuropsychologist at the Institute for Frontier Areas of Psychology and Mental Health in Germany, is the author of many papers on time perception and the book

* A study in 2013 found that synesthesia occurs more often in autistic people than the general population and is associated with the high serotonin levels present in about 25 percent of autistic people. Perhaps psilocin's excitation of serotonin receptors is, in the case of synesthesia anyway, analogous to naturally high serotonin levels in the brain. See: S. Baron-Cohen, D. Johnson, J. Asher, et al., "Is Synesthesia More Common in Autism?," *Molecular Autism* 4 (2013): 40.

Altered States of Consciousness: Experiences Out of Time and Self. He and his colleagues have documented that time perception is closely affiliated with mood states like boredom, stress, or excitement. You have experienced this before. When you are bored by the same old, same old, time drags. When you are stimulated by something new, time flies. That's because our brains don't spend capital recording redundant memories. A widely reproduced finding showed that adults, regardless of culture, perceive time as passing faster the older they get. This may be because "based on prior experience, the brain constantly makes predictions about what might happen next."

Those predictions become increasingly frequent as we get older and develop more routines, leading to "a reduced autobiographical memory load." The brain doesn't rewrite predicted memories, suggested David Eagleman, who studies the neural mechanisms of time perception. "When you are a child, and everything is novel, the richness of the memory gives the impression of increased time passage—for example, when looking back at the end of a childhood summer." But for adults, summer seems to pass faster and faster. However, a tripping adult brain records experiences similarly to a child's.

Time may slow down for the tripper because psilocin acts on the serotonin system, and the genes that regulate the serotonin system are also related to time perception. At the beginning of a trip, Dr. Wittmann told me, "you have a state of hyper-mindfulness and time slows down; looking back, because you have so many memories of what happened, you think the trip lasted very long. But in between, at least with a high dose, you may also lose the sense of time passage altogether."

When we travel, maybe the days seem so full and long because our brains are busy recording new experiences. Similarly, a psychedelic trip really *is* like taking a trip. Both experience types present the opportunity for novel perceptions that our brains expend energy recording (or, to look at it another way, both experiences upend our brains' predictive edge), and both operate on serotonin pathways. Indeed, after a psychedelic trip, my brain feels like it does after I've spent the day sightseeing and wandering around museums.

Insights and Revelations

During a trip, one can drift in and out of awareness, from a trancelike state to a lucid one. I once popped out of a trance to realize I'd been staring at the sun (disconcerting but no harm done). You may feel increased empathy for others, including animals and natural systems, even for yourself. For many weeks, I sat in on meetings of the Autistic Psychedelic Community. I learned that one of the reasons why folks in the society take psilocybin is to absolve themselves. Sometimes when an autistic person expresses their true self, they are socially rejected, and that damages their self-confidence. It is, as one person in the group described it, crushing. The drug helps them feel empathy toward themselves and to find self-acceptance.

You may experience insights into how you interact with others, society, and nature, and those insights may come with a powerful sense of truth, as if you knew this truth all along but never realized it before. "The predominant feeling during a session," wrote James Fadiman, a longtime researcher into psychedelics, "is not of discovering something new, alien, or foreign, but of recalling and reuniting with an unassailable clarity that has been latent in one's own mind."

These insights can have lasting benefits. A scuba diver instructor I know ate a mushroom-infused chocolate when he was eighteen. He didn't know how much psilocybin was in it, but soon realized he needed to be in a safe place and so decided to drive home. On the drive, the drug came on—strong. Buildings on the street stretched and grew, folding over, darkening the sun. He became panicky. And then he realized, "I have two paths here. One is to succumb to the panic, wreck the car, maybe go to the hospital. The other [is] to stay calm and get home safely." He decided to calm himself, and he did. "I realized it is in those moments when the body experiences panic that I need to keep my head the most, and that has had a ripple effect in my life. Panic is a really bad thing when someone is deep in the water. But I know how to sit with whatever is coming up for a client; I know I can choose to use my head and take the calm path, and I help them do the same."

Tripping creates a kind of hybrid consciousness where I am aware but in a fantasy. It's a state of metacognition, which sounds like a superpower but is simply thinking about how I'm thinking. I can objectively observe myself having internal dialogues. The drug allows me to see right through me and, subsequently, realize why I do some of the things I do.

But the opposite can happen, too. It's possible to misunderstand what you are feeling, which can lead to anxiety. For example, becoming super conscious of your breathing might lead to hyperventilation or feeling like you are suffering an asthma attack. This happened to a fellow I know. He only started to feel better when his facilitator coached him through slow breathing. During one trip, I had to pee, but in my mind, I pathologized the feeling and became convinced I had a bladder infection. (I didn't.) Under a strong dose, or a dose that is strong for you, it's possible to even experience memory blanks and not recognize where you are, who you are with, or even who you are. One woman who experienced this told me, "All of a sudden, I didn't know who I was or who my boyfriend was or where we were. And I just remember being so worried that I was going to be stuck in that state forever and end up locked in a psychiatric asylum." But less strong doses, like those I've preferred, allowed me to do everything I needed to do. I might gaze, trancelike, at the landscape for a while, but when I felt the urge to pee, I could pull myself together to go find a private bush.

What so many people who experiment with magic mushrooms are hoping for are personal revelations, the ten-years-of-therapy-in-six-hours effect. Hypothetically, during the course of a trip the parts of the brain involved in ego—the DMN is generally considered the culprit—are temporarily disengaged or differently engaged. Perhaps this leads to a breakdown of the ego constructs we've built over the course of our lives to excuse or explain away behaviors, including long-established opinions and judgments. And free of those ego constructs, some trippers gain insights into themselves. For example, on an afternoon trip with a large group of middle-aged people, one man dredged up a memory about a soured business deal, and he told me he realized he'd done a partner wrong. He'd kept that inside for a long time. I know what it feels like to carry guilt but be unable, for whatever reason, to resolve it. I've spent plenty of time justifying past actions in lieu of resolving some guilty feeling, a kind of "I did this because he did that" legitimizing of my behavior. I have a portfolio of such episodes from my life. But what the drug seems able to do is help you set aside the narrative that stops you from seeing what the real cause of a problem is. And that can feel like a revelation.

Insights that happen on trips don't always have to do with personal issues. Sometimes they can be about work or an aid to solving problems. A special education teacher told me about a mushroom trip that helped her reach one of her students in a new and, it turns out, effective way. "He was clearly delayed, but also distant, disgruntled, and often distraught. He spent much of his time in class hiding under his desk. We were very concerned, and nothing seemed to help." One evening, she and some other faculty gathered to take mushrooms. "The bulk of the evening was spent in uproarious laughter, which was something we needed." But toward the end, work life inevitably crept into their conversation, "and I asked my colleagues if they thought this puzzling child just needed space to play and work through some of the emotional burdens that seemed to be hindering his progress." That moment of clarity led to her switching the child from classroom work to a social-emotional play group for a little while. "I had the child start coming in with a few other students three days a week to play different cooperative games, to draw, and to play with a dollhouse together, and the effects were astounding. He is a gifted artist and really shined during our drawing sessions and even started teaching the other kids a few of his skills. His self-confidence grew, he had buddies, and felt more at home. After six weeks, he was ready to go back to working on his academic goals, and he was like a new kid."

Trippers may experience many such insights at any time during a trip, and those insights are often coupled with an indubitable sense that they are true. While insights and revelations are ultimately personal, there is one that seems to be widely shared: an incontrovertible awareness that we are connected to everything in nature. This insight is often associated with ego dissolution and the transpersonal state.

Ego Dissolution

Trippers often report a sense that everything around them is alive; obviously, trees are alive, but do we regularly sense their *life*? Once during a trip, I sensed the landscape respiring, breathing in and out. I started to breathe with the landscape, too, and I felt a distinct and lovely sensation that everything I could see was as alive as I was, and all this life was in a kind of glorious syncopation. Some trippers interpret this feeling of connectedness with nature as a mystical experience, the result, perhaps, of so-called ego dissolution (a.k.a. ego death, ego loss, ego disintegration), a loss of subjective self-identity. The authors of the paper "Ego-Dissolution and Psychedelics" describe it as an experience of "depersonalization and derealization." It can feel very good or very bad; something welcomed or something fought against.

Ego dissolution is likelier to occur with large doses, and some people take large doses because they want to experience it. It can run the gamut from blissful to terrifying, from nature relatedness to experiencing one's own death and rebirth, even experiences of multidimensionality where one perceives everything is everywhere at once or, conversely, absolute nothingness.

There are many ways to describe ego dissolution. I like one from Manesh Girn, a neuroscientist who hosts the Psychedelic Scientist YouTube channel. He described the ego as a combination of the embodied self, which is one's sense of self-experience, what we experience with our bodies, and the autobiographical self, which is related to the narratives we tell ourselves about ourselves, our identities, our life stories. In a psychedelic trip, those self-identities may dissolve. Our autobiographical selves lose track of how we describe who we are, and our embodied selves lose their boundaries and merge with the world around us so that our sense of being distinct from nature melts. "We are left," he said, "with this unitive consciousness, where there is undifferentiated awareness of all that is occurring without feeling like we are a distinct person experiencing it."

Coming Down

You will probably know when you have emerged from the strongest part of the trip. With the cessation of effects, I've felt a growing sense that my old self is back. It felt safe and yet slightly disappointing, like I was on my way home after a big adventure. Others have told me they've felt relief that the ordeal was almost over. You may continue to have significant personal insights once your trip ends, but without the distraction of hallucinatory or dissociative states. Coming down from a trip can be euphoric, with welling feelings of emotion, love, and gratitude, or relief, even if the trip wasn't stressful. It's a time when I write my husband love texts. I seem to recall blowing kisses of thanks to various trip sitters. I always cry: happy, overwhelmed tears like a child does when they receive a birthday gift they've yearned for. But then I've never had a bad trip.

Barring those who have devastating trips where their fears or worries linger, or they have caused harm to themselves or others during the trip, most people do not experience aftereffects, although some may feel worn-out or depressed, have a headache, or suffer sleep issues, from night terrors to insomnia. Coming down is often associated with body sensations; I've felt the desire to stretch, to sit up and look around anew, and waves of cozy feelings, as if I am snuggling under the covers. You may feel sensitive or vulnerable, or compelled to share your experience right away. After a trip, I always feel introverted. At the conclusion of a retreat experience where I had tripped with a group in a yoga studio, a woman from Long Island plunked herself at the end of my mat to unload, but I couldn't follow and was so totally inarticulate I could only smile and nod and let out an occasional "Wow!" when she paused for my response.

It might take you days or months after a trip to process any insights; it is not uncommon for those insights to reappear in your mind with greater clarity as time goes on. At a ceremony held by a Mazatec shaman, we were advised to avoid alcohol, sex, and extravagant financial expenditures for four days after our trip, to keep the experience introspective and give ourselves time to digest. Nor is it uncommon to recall your trip vividly for years afterward. I distinctly remember the details of trips, which is good because the notes I wrote are utterly unintelligible.

In some cases, trippers who have a neurological event like blacking out weeks after the trip can experience intense worry that their trip was somehow responsible. I've talked to a few people who have experienced this, and it is a torment, but it is also difficult to prove psilocybin was the cause. One fellow I know had a stressful trip and two weeks later drove a rental pickup truck across two lanes of oncoming traffic without speeding up, braking, or swerving, and into someone's living room. He "came to" after hearing sirens to find himself sitting in a demolished truck with three airbags deployed. He spent the next eighteen months trying to figure out what happened, enduring tests by neurologists, seeking advice from traditional Chinese medicine doctors and psychotherapists who eventually diagnosed the event as disassociation with amnesia. But whether that was a result of taking the mushroom, he can't know. "Sometimes," he wrote me, "things come together that cannot be anticipated or explained."

It may not be surprising to hear that many people are in no rush to trip again, and some never do. Others may feel they've got a grip on the experience after a first trip, only to find the next time around, they did not. "I got slapped," one woman told me. "It's like the mushrooms said, 'Nope. You *don't* got this.'"

People trip in all kinds of patterns: I am affiliated with an annual tripping group whose participants change slightly from year to year. I know folks who have tripped several times in a short period and then called it quits, and others who trip the way some people drink martinis. However frequently the experience is approached, according to the "philosophical entertainer" Alan Watts, "if you get the message, hang up the phone. For psychedelic drugs are simply instruments, like microscopes, telescopes, and telephones. The biologist does not sit with eye permanently glued to the microscope, he goes away and works on what he has seen."

Is Psilocybin Therapeutic?

The short answer is it can be, though how is far from being totally understood. Researchers have associated psilocybin-induced ego dissolution with decreases in depression, anxiety, and existential distress for some people. The prevailing hypothesis is that ego dissolution occurs when the default mode network is downregulated, as it seems to be during a trip, and since the DMN is strongly related to the narrative self, said Dr. Wittmann, "the downregulation of the DMN or the decrease of DMN interaction with the rest of the brain would be related to a reduced narrative self. This could be a decisive step towards a total ego dissolution."

Personal insights that arrive with deep certainty may arise from psilocybin-induced ego dissolution. Peter Hendricks is a gregarious and jolly family man who studies psilocybin and a range of addiction, anxiety, and other disorders at the University of Alabama at Birmingham. He explained to me that most if not all mental health problems may be understood as a form of "pathological self-focus." Not necessarily narcissistic, "but one's daily experience becomes highly focused on the self." So, the addicted person may be fixated on using, the chronic pain sufferer is fixated on their pain, the person dealing with an end-of-life prognosis is fixated on their mortality. "When we ingest psilocybin," Dr. Hendricks said, "it promotes chaotic, hyper-associative cognition. Your field of awareness opens. That's when people have moments of awe, and you aren't thinking about yourself anymore." You become small relative to your expanded sense of reality, and "liberated from those nagging self-directed thoughts, you are able to make the connections we call insights." Insights that may be therapeutic.

For some, ego dissolution is a spiritual experience, and that can have therapeutic ramifications as well. The paper "Psilocybin Produces Substantial and Sustained Decreases in Depression and Anxiety in Patients with Life-Threatening Cancer: A Randomized Double-Blind Trial" reported on patients who answered questions about how personally meaningful the trip was, how spiritually significant, and, five weeks later, how depressed or anxious they felt. The study showed that the more spiritually significant and meaningful the trip was, the less anxiety and depression the patient reported. Other studies have shown psilocybin-induced mystical experiences, when combined with the patient's own spirituality, produced positive changes in their attitudes, including how they relate to others.

Other Ways Psilocybin May Be Therapeutic

Another factor that may explain psilocybin's potential efficacy as a therapeutic drug is psilocybin-induced neuroplasticity, the ability of the brain to "strengthen, loosen, and reorganize its connections." Psilocybin may change serotonergic pathways that have been altered by traumatic experience, like those some veterans have encountered. It may do this by inducing an acute "hyperplastic" state where the brain learns new things and then, after the trip is over, it leaves open a "window of plasticity," where the new knowledge may be set, leading to "positive neuroplasticity," or serotonergic pathways that are symptom-free.

You've seen examples of neuroplasticity. Within forty-eight hours, a newly hatched duckling will imprint on its mother duck—or the family dog, if that is who is there; afterward, the window of attachment closes. Children seem able to learn languages more readily than adults: that lasts until they are five or six years old, and thereafter languages are harder to learn. These are called *critical periods*, periods of time when the brain is open to certain kinds of imprinting. It could be that psychedelics open critical periods of their own. "A radical shift in your environment may open a critical period," explained Gül Dölen at the Horizons 2022 psychedelic conference, like a stroke or a psychedelic experience. Her work as a psychologist at Johns Hopkins University suggests that not only do psychedelics seem to open critical periods but different psychedelics keep them open for different lengths of time. For psilocybin, she explained, that period lasts for about two weeks after the trip—a period when, if coupled with therapy, people may experience changes in the structure or function of neural pathways that could relieve them of problematic symptoms.

But this isn't always what happens. Sometimes the drug seems to reinforce problematic serotonergic pathways. I have spoken with a few folks whose negative self-talk acquired added agency during the trip. "Yes," the mushrooms seemed to say, "you *are* an awful person." The psychological risks are real. "It's a powerful experience and without crucial integration in a non-drug setting, it can be disturbing and ask more questions than are answered," said Ben Sessa, a charismatic psychiatrist and psychedelic researcher from the UK. He believes therapy can play an important role in helping patients integrate their experience and learn from it.

Or in the case of a bad trip, recover from it.

3

Bad Trips

A dust storm roared outside the Zendo tent at the Burning Man festival. The tent sides were flapping, and inside, white dust like blackboard chalk swirled and settled over everything: eyelashes and coffee cups and clipboards. Two rangers, part of a volunteer patrol that helps people who express a need for help, emerged from the whiteout, ushering in a very large seminaked tattooed man in a floppy woman's sun hat. He was disoriented and lurched about, like a slow-moving eighteen-wheeler without a driver.

I wasn't in the tent because I was having a bad trip. I wasn't even tripping. In general, I found the festival so stimulating, with its dust storms and nighttime blinky lights, that I didn't feel I could handle any psychedelic amplification. I've never had a bad trip, but I wanted to see how Zendo managed them.

This was Zendo Project's tenth year of providing psychedelic peer support, and during the weeklong festival, they ministered to more than six hundred people. *Zendo* means "a place for Zen meditation," but from what I could tell, this Zendo was a place where confused and overwhelmed people without obvious health issues could lie down for a while and, if necessary, continue their trip in the safety of a swamp-cooled yurt.

A volunteer stepped up and asked the fellow in the floppy hat what drugs he'd taken, and he replied, "The doctor's dose."

Not particularly helpful, but that's what the Zendo folks regularly deal with. A medical professional on staff checked his blood pressure and for signs of dehydration—the most common medical emergency on the playa, followed by costume-based rashes—and since he showed no physical distress but rather a kind of zombielike confusion, he was asked if he would like to lie down. When he said yes, he was led to one of the yurts in the camp.* I peeked in. It was comfortably cool and humid, and relatively dust-free (meaning you didn't need goggles). Eight trippers lay on yoga mats, and next to each one sat a volunteer, some talking, others holding a hand, but most just providing silent company. It was a reminder to me that a trip can go downhill if the environment where you are tripping suddenly seems threatening. Maybe the floppy-hatted guy had been happily boogying in the desert, but when those dust storms rolled in and over him, massive and illimitable, he couldn't handle it. In the comfort of the yurt, and in the company of sober trip sitters, the people I observed were calm, their trips continuing inward.

Bad trips are common; they just aren't commonly described. Based on three controlled studies, about 30 percent of healthy people receiving a high dose of psilocybin experienced significant anxiety and fear. Trips can transpire on a spectrum of badness, and for any number of reasons, from scary settings like a dust storm or creepy people around you to physical problems like temporary paralysis or your own reaction to upsetting thoughts and images that arise from your subconscious. The twinges of anxiety that so often occur when the drug is coming

* In 2017, a young woman who had attended a California music festival died not long after Zendo volunteers had tried to help her through what they thought was a bad trip. While Zendo can help trippers, it is not a medical facility.

on can trigger a bad trip if not assuaged. Sean O'Carroll, an Australian psychotherapist who has treated hundreds of people with post-trip issues, told me that a bad trip often begins when a tripper responds with fear and resistance to those first feelings of their mind becoming loosened from its moorings.

Of course, what constitutes a bad trip is subjective, and folks in the psychedelic community often say there are no bad trips, only challenging trips. But I don't think that's true, and what's more, I think that point of view dismisses how tough a time some people have. Maybe a trip can be termed *challenging* if the difficult, horrible, or confusing experience you had resolves by the time you sober up. But a bad trip that stays with you, that in some way shakes your world, is traumatic, even causing a kind of PTSD that can disrupt your daily life afterward. For those who continue to feel distress after the trip is over, the consensus is that therapy should be sought.

What Is a Bad Trip Like?

The clinical literature describes bad trips as experiences that include fear or panic, paranoia, sadness or depressed mood, anger, confusion, and dissociation. Laypeople describe bad trips as experiences that include encountering terrifying entities and places, experiencing anguish, grief, and despair, revisiting traumatic childhood experiences, having painful insights, upsetting realizations, or an anxiety attack, and thinking you have gone mad. Even a nonexperience can be a bad trip. Folks hoping to have a mystical experience but don't may question their worthiness. For others, a bad trip is characterized by intense frustration from failure to "go deep."

Some bad trips fade once the psilocybin is out of your system. Others leave the tripper disturbed—sometimes for a long time. Sean O'Carroll has treated many people who have suffered from "non-ordinary state trauma," the trauma caused by a bad trip, and it can leave someone in a "very wobbly or fragile place." His framework for understanding and helping these folks is informed by data he's gleaned from hundreds of firsthand client experiences. "In each case," he said in a conversation with me, "I'd hear about the trip experience, their ongoing difficulties, and the context in which the trip took place." After a decade of listening to these stories, he noticed bad trips tend to cluster into categories.

The severest category he has observed includes "existential" themes relating to the person's fundamental sense of themselves or reality. "In these instances, the integrity of the self is often felt to be fragile or under persistent threat," he said. In an existentially themed bad trip, the individual resists the drift toward ego dissolution, that aspect of a trip where the boundary between the self and the world disintegrates. The tension between the psychedelic's pull toward dissolution and the ego's struggle to maintain its integrity can lead to persistent feelings of fragmentation—the sense that nothing is real, that no one else really exists. The tripper may think, *I am fundamentally alone, I am not me.* That's in line with data collected by the Challenging Psychedelic Experiences Project, which quantified the experiences of over six hundred people who reported extended difficulties following a trip.

They found the most common forms of extended difficulty—meaning difficulties that persist after the trip is over—were anxiety and panic attacks, existential struggle, social disconnection, depersonalization, and potentially crippling derealization.

Derealization is when people can't tell if they are in a dream or not. A heartbreaking case of this may have occurred in 2023, when an emotionally distressed off-duty pilot tried to down a passenger plane two days after a mushroom trip and having had no sleep for forty hours, because he thought he was dreaming and just wanted to wake up. People in a fragile state of mind because of psychedelic use might be best off seeking care from a therapist with a background in psychedelics, because to an everyday psychologist, these symptoms suggest psychosis.

Many observers have commented on bad trip archetypes. For example, dying—entering the void, falling into the abyss, being overpowered—is common bad trip territory. When I solicited stories from my social media circle, they described encounters with the pits of hell, black holes, holes gooey with rot, and vortexes that they are either sucked into or, despite the terror, drawn to. In a talk about his book *Sacred Knowledge*, the eminent psychedelic researcher and psychologist William Richards told a story about someone who, when his trip finally subsided, said, "Phew. I made it," and when Richards asked what he meant, the tripper explained he had spent his entire time swimming against a vortex, trying desperately not to be sucked under. Another fellow "dreamed" he was lying on a beach and could not move. As the tide slowly came in, he still couldn't move. Even after he was covered with water and his breath began to gurgle, he could not move. He remained aware as the ocean covered him more and more deeply.

Another archetype is the threatening encounter. A tripper may deal with demons, monsters, parasites, spiders, and, in one case, a giant starved crow, which attempt to threaten, consume, or possess the tripper's soul, or point out that he is a bad person. In the cases where the tripper continues to feel haunted after the drug has worn off, therapy may help put his experience into some kind of manageable perspective. For example, some trippers come around to see their monsters as personifications of their vanity, envy, or shame. "What is so important here," wrote Dr. Richards in *Sacred Knowledge*, "is the discovery that the monster has meaning and in itself is an invitation to enhanced psychological health and spiritual maturation. Its purpose is not to torment but to teach." This explains, I think, the evolution of the nomenclature. Yesterday's bad trip is today's enlightening one.

For many people, magic mushrooms can dredge up memories. But sometimes the memory is not familiar; it may be a false memory or a repressed memory—though repressed memories are, according to the American Psychological Association, extremely rare. This can create great uncertainty and mental anguish, as when someone experiences a "memory" of child sexual abuse. When people are confronted with disturbing or confusing scenarios in their trips, it can be difficult to differentiate between memory and symbolic visions, said Sean O'Carroll, and they may wish to seek help.

More garden-variety bad trips are those where you spend six hours experiencing anxiety and paranoia. Elizabeth and her husband were married in a farm setting in the presence of family and friends. On their way to the reception, they each ate a little mushroom a friend had given them. No big deal, they thought, but an hour later, Elizabeth couldn't make sense of anything around her. What's more, the reception was held at outdoor tables with wildflower arrangements and candles, but the day was windy, and the candles kept blowing over and setting the centerpieces on fire—distracting enough for a sober person. "It was *disconcerting*," she said. "We were the focus and

I realized, *I'm on a trip and everyone is going to know*, and I felt something bad was going to happen, that I was a bad girl. For a while, I didn't know what was going on. The anxiety came from the expectation to function in the world, which, of course, I was unable to do."

We are, as the social neurologist John Cacioppo described, obligatorily gregarious. It is a biological imperative to function socially. "If I encountered a tiger by myself, I might get eaten," said Elizabeth. "But if I encounter a tiger with ten other people, I'd have a good chance. So, feeling isolated and alienated is deeply unnerving."

Bad Trip Triggers

Tripping can amplify your personality and mood; psychedelics are, as Stanislav Grof has said, "non-specific amplifiers," so someone in the midst of an ugly divorce or in the throes of a nervous breakdown may find a psychedelic trip makes them feel worse. That doesn't mean you have to be in a completely clear and peaceful head to trip; many people consume magic mushrooms *because* they feel distress. But your mood at the time, your reasons for wanting to trip, and other drugs you might be taking all can influence the nature of the experience.

There are certainly some folks who should avoid psychedelics. People with a personal or family history of bipolar disorder or schizophrenia have long been barred from clinical trials and psychedelic retreats for fear the drug could trigger a psychotic episode. The psychedelic community also warns that mushrooms (and LSD and even pot) may trigger psychosis that leads to hospitalizations. One report I read in the Shroomery forum told of a fellow who didn't know he had a predisposition for schizophrenia, took mushrooms, and flipped out, believing he was God and had to kill the devil, who happened to be his friend. He armed himself with a knife and was within a half mile of his friend's house when he realized what he was about to do. He ended up being hospitalized for four months.

If you have a diagnosis of schizophrenia or bipolar disorder, that's one thing. Psilocybin may be risky for you. But what if you have a genetic disposition to these disorders and don't know it? That's a risk, too. And even then, mental diagnoses aren't always definitive. My mother was diagnosed with depression, then bipolar, then depression again. Should I have avoided psychedelics? The bottom line is that everyone must do their own risk assessment. Some companies are stepping into that void by offering pharmacogenomic tests purporting to determine how your DNA profile might affect your response. I wouldn't place much stock in those tests, at least not yet. There is simply too much that is unknown about the genetics of mental illness. But in the future, there may be a dependable test to determine if you are prone to a psychotic break while under the influence of psilocybin.

A trip shares characteristics with psychosis: in both cases, you can lose track of normal reality. When LSD was discovered, it was originally called a *psychomimetic*, because its effects resemble psychotic symptoms. But the term was replaced with *psychedelic*, derived from the Greek meaning "mind manifesting," which makes room for the variety of experiences a tripper may have. Anyway, psychedelics rarely trigger psychotic episodes in healthy people. And even the prohibition for bipolar users may change in the future. Emerging research suggests psychedelics might be safe for people with bipolar disorder, even act as a potential treatment, but the science is nascent.

Trips can have episodes of badness or just be tough the whole way through, and they are triggered by a vast array of causes, from the trite, like stepping in dog poop, to the profound, like experiencing mushroom-induced paralysis. A trip that is otherwise going fine can be sidetracked by an upsetting encounter, like being touched inappropriately by a guide or bullied by a drunk person at a music festival, as happened to Aaron Hodgins Davis, founder of the mushroom farm Hodgins Harvest. Once, tripping at a music festival, he got separated from his friends and then couldn't find his tent:

> I was walking down a main drag with bright spotlights overhead. Two guys without shirts came running down the street, they were all hyped up on something, roughhousing and yelling loudly. One of them looked directly at me and then intentionally ran into me, knocking me back a few steps. He let out an abrasive laugh right in my face and ran off into the night. That interaction started me down a psychological spiral. It made me question the intentions of the people around me and made me feel unsafe. My confusion over where I was turned into a desperation to find my friends. I searched the festival grounds for what felt like hours, to no avail. I was wandering through a field of tents with a hedgerow to one side, far away from the bright lights of the main drag and finally decided to lay down in the woods to try and sleep. I stumbled into the hedgerow and got tangled in some undergrowth, and fell and then just laid there, giving up. In my paranoid state, I remember thinking there were poisonous snakes all around

> me and if I went to sleep, they would bite me. I closed my eyes and visualized bulging, bleeding eyeballs popping out of people's heads. I resigned myself to what I thought was certain death and fell asleep. I woke up a few hours later with the sun out and the Tennessee heat starting to set in for the day. I was lying in a pricker bush and still very much alive.

During my afternoon at Zendo, I observed one fellow, an Asian man in a dusty top hat, emerge from the yurt, ready to go back onto the playa. "Don't eat any more shady danishes," called one of the volunteers as he headed into the whiteout. The fellow had been dosed: he'd consumed a pastry spiked with psychedelics. Dosing an unsuspecting person (or animal) is cruel, dangerous, and a fourth-degree felony. Despite all that, it is not uncommon, and it can absolutely lead to a bad trip. I saw it happen at Burning Man, where one of our camp members was given a chocolate chip mini-muffin. "Within twenty minutes, I started feeling off," said my fellow camper. "Dizzy. And my upper body really relaxed. I thought, *What is this?* and then, *Oh shit! The muffin.*" Shortly after, she lost all muscle strength and was unable to stand. "I got really scared. I went into this fear space. *I don't know what's in this. Am I going to die?*" She needed to be carefully nursed during the duration of whatever drug it was. Everything turned out all right, but the obvious prevention method is to avoid taking any food, drink, smoke, bumps of powders, or mini-muffins from a stranger.*

Not only was my campmate dosed, but she may also have experienced wood lovers' paralysis (also wood lover, wood lov-

* If you do get dosed, the only way to find out what you were dosed with is to get a toxicology report from a hospital as soon after the experience as possible.

er's, wood-lovers, or wood-lover paralysis). Wood-decomposing psychedelic mushrooms (versus *P. cubensis*, which grows on manure), like *P. subaeruginosa*, *P. cyanescens*, or *P. azurescens*, can cause paralysis that is characterized by a sudden loss of muscle strength, falling, and, to varying degrees, not being able to get back up.* It doesn't seem to matter if you eat them fresh or dried, or where they grow. It is maddeningly random. Sometimes the weakness occurs in waves, lasting minutes to an hour. It mainly affects the legs and hands, but can affect the lips, swallowing, and breathing. In a 2020 survey conducted by Dr. Symon Beck and mycologist Caine Barlow in (mostly) Australia, they found that of four hundred respondents, 40 percent suffered the paralysis at least once (that number is high due to recruitment bias—the authors were seeking individuals who had experienced WLP). In most cases, the paralysis occurs within the first two hours of the trip. There usually aren't any residual effects, unless, of course, you hit your head when you collapsed, or you collapsed in a dangerous setting, like in freezing temperatures outside or while swimming, or in the case of one person who fell and was paralyzed next to an electric heater, severe burns.

Wood lovers' paralysis is not widely recognized by hospital personnel. In the comments section of a recorded lecture by Dr. Beck and Barlow, one man recounted experiencing paralysis of his legs and arms, but then having trouble breathing when he lay down. He called an ambulance, which took him to the hospital, but the staff just thought he was high on drugs and should go home. "The last thing I saw was the doctor's shoe.

* It's because of wood lovers' paralysis that the state of Oregon has determined only *P. cubensis*, which is not implicated in the paralysis, is acceptable in its state-sanctioned psilocybin-assisted therapy programs.

My heart stopped beating. I was dead. They told me afterward that they tried to reanimate me for one minute. I was four days in the hospital. Everything is fine now. I still love shrooms. I've learned a lot from them, but also, they killed me."

This story, if true, is an extreme example, but not out of the realm of possibility. In an email exchange, Barlow suggested that WLP could indeed impact respiration "in severe cases," or if the mushroom is taken in combination with other substances or medicines that impair breathing. (Opioids come to mind as a potentially risky drug to combine with wood-loving *Psilocybe*.)

No one knows for sure why or how wood lovers' paralysis happens—there is nothing in the medical literature regarding the phenomenon so far—but one hypothesis credits the effects on dephosphorylated aeruginascin, a psilocybin derivative present in some species. It does not cross the blood-brain barrier but lingers in the peripheral nervous system, in which case, proposed Dr. Beck, it may act like an antinicotinic neuromuscular blocker. Should a tripper experience wood lovers' paralysis and go to the hospital, it is important to reveal that mushrooms have been consumed because the paralysis can be misdiagnosed as the autoimmune disorder myasthenia gravis, and the tripper might end up intubated—a sure way to make a bad trip worse.

A trip to the hospital is not the time to lie about taking magic mushrooms because you are afraid of arrest, though it's understandable. Paranoia and fear of arrest are common bad trip triggers and one of the main reasons why people seek haven in the harm-reduction tents at music festivals, said Tobey Tobey, a harm-reduction specialist and founder of Altered States Integration. It is particularly acute for African Americans, many of whom have experienced the war on drugs as a war on Black people. Here's a scenario they face: a tripping Black person may become paranoid about getting arrested, leading to a bad trip and associated frantic behaviors, which in turn may lead to his being arrested and tossed into jail. Some metropolitan areas are looking to educate their first responders about psychedelics. For example, Denver, which has legalized psilocybin, is on track to develop a training initiative that would, in essence, replace cuffs with calming words. I heard Joseph Montoya of the Denver Police Department speak at the Psychedelic Science conference in 2023, and he said that while police face many challenges, "none of them are psilocybin." Still, it's important to check the standing of the laws in your state. Wikipedia's "Legal Status of Psilocybin Mushrooms" seems to provide current info on the laws in US cities and states and in foreign countries.

What surprised me the most about bad trips is they happen in clinical trials, too, where not only is it legal to be tripping but usually you are surrounded with people who are focused on your well-being the entire time. Bad trip triggers in clinical trials may bubble up from inside the tripper or reflect how the tripper relates to the researchers. And they do tend to be underreported. That may be because participants fill out questionnaires after their trip, but the questions don't always align well with their experience. You can have a bad trip that is deeply upsetting for long afterward, but did you experience toxic effects, complications, harms? These are terms that have been used on post-trip questionnaires, but they just don't apply to every bad trip. Additionally, how does the tripper respond when aspects of a bad trip turn out to have therapeutic value? What if their trip was only bad at one point in time, but otherwise was okay? It is very difficult to come up with questions that might address these variables. And yet, if the right questions are not asked, then bad trips might not be reported.

Participants may also underreport their negative experiences if studies rely on them to bring the subject up. Bad trips that haunt folks for days or months afterward don't necessarily get included in the trial's data summary due to follow-up constraints, and unresolved trips can lead to increased depression or anxiety way after the trip is over, which may be discounted because they cannot be definitively attributed to the trip.

All this has contributed to an overly rosy picture of psychedelic medicine. Indeed, in the case of that pilot who, two days after a psychedelic trip, tried to down a passenger plane, various experts in the field came out to say it is very unlikely the psilocybin had anything to do with it. I'm not so sure. I think transparency—and humility—is important not only for the evolution of public health but also for building public trust. "A serious effort to examine bad trips can be perceived as positioning oneself 'against' the movement," wrote Rachael Petersen in the *Harvard Divinity Bulletin*. "This, I think, is a shame, because a full account of *what* psychedelics are, *how* they are, and—most vexingly—*why* they are, is as beholden to the harrowing as it is to the heavenly, to abject terror and to unbearable bliss."

Only by thoroughly and consistently reporting our bad trips *as they seem to us*, either in the forums or by participating in projects like the Challenging Psychedelic Experiences Project, will knowledge about bad trips grow. With enough anecdotal stories about bad trips in hand, researchers might see patterns and trends that could help us better understand why they happen and how to support people when they do. Indeed, I hope that as psilocybin is decriminalized and legalized in cities and states across the country, local regulators will make every effort to ensure that vulnerable people—like first-time users or mentally fragile folks—know the risks of bad trips and how to get therapy should they need it.

Trip Killers and Support Systems

If someone is in the midst of a bad trip, they or their scared friends may decide to go to a hospital. "Mainly, we see people who have done multiple psychedelics at once," said Linda Johnson, a retired ER nurse who worked in a rural Colorado hospital for thirty-three years. "I remember one guy who was hallucinating and manic. He was talking to people that weren't there and pointing to things in the air. He wouldn't stay in the bed, and we had to keep him in his room. It was a pain in the butt, really. We eventually gave him Lorazepam [a benzodiazepine]." A person trapped in the panicky spiral of a bad trip will often be given benzodiazepine as a trip killer, though it doesn't kill the trip, nor does it discontinue the visuals. But it can calm a tripper down.

The key word here is *calming*, and there are a variety of ways people get there. Some guides use MDMA to modify their client's mushroom trip should it become overwhelming, a practice called *hippie flipping*. (It's not as crazy as it sounds: a study in 2023 suggested the co-use of MDMA with psilocybin could "buffer against challenging experiences and enhance positive experiences.") There are all kinds of individual prescriptions to calm distressed trippers. Some suggest warm showers. One trip sitter said he found holding someone's hands, looking deeply into their eyes, and breathing slowly with them "works wonders." Another suggested offering a lollipop because sucking on it can redirect the mind. Plus, "it's cheap." One friend told me he keeps a Xanax (a benzodiazepine) and a pot of chamomile tea on hand to give to someone, or take himself, in case of a bad trip. People who chug a bottle of vodka to calm down always seem to be sorry they did. Some head shops sell trip killers, like Trip Stopper 3000, a combination of valerian extract and maltodextrin. (If Trip Stopper works, it may be a great example of the power of placebo.) Even though we don't talk about bad trips much in the psychedelic world, industry knows they exist; otherwise, why would the psychedelic pharmaceutical company MindMed file a patent for an LSD "neutralizer technology"? But the anecdotal consensus is if you want to end a trip, probably the best bet is an antipsychotic like Thorazine or olanzapine.

The idea of developing an antidote to the mushroom experience is indicative of the way we engage with drug side effects: we use another drug. Which begs the question: What do indigenous people do when one of their community members has a bad trip? It seems natural to look for a model among people with long experience of use. But when I spoke with Mario Alonso Martínez Cordero, who has studied indigenous mushroom rituals, he said, "That's a difficult topic." Some healers, he said, think foreigners who come to Mexico to trip but fail to follow the ceremony's rigorous protocols can have bad experiences. But "when you ask healers about bad trips, they will not admit to any difficult experiences." It may be their trips are mitigated by the collective ritual of the ceremony. Maybe the potential challenges of psychedelics are part of their understanding of the mushroom's virtue.

In its own way, the psychedelic community in the US and Europe has always sought to protect trippers from harm. This sense of social responsibility—altered states responsibility anyway—is one of the community's defining aspects. Groups like Family

Security in Denver provided calm spaces at music events for freaked-out trippers in the 1960s. The Hog Farm helped out at Woodstock. The Brew Ha Ha at Rainbow Gatherings and the Zendo Project at the Burning Man festival provide similar functions today. In Europe, the Kosmicare project supports trippers at the Boom Festival in Portugal, and PsyCare does the same in the UK. The Loop, also in the UK, caters to the rave scene and texts followers warnings like, "Donald Trump pills [featuring a caricature of the former president's face on a tab of MDMA] causing serious medical problems across the UK this weekend."

Event organizers may not be legally required to provide psychedelic support, but these organizations are welcome because they provide an important service for panicky or paranoid people who might otherwise get themselves in trouble. For example, years ago, Tobey Tobey lost his companion at a music festival and became convinced he was receiving SOS messages from her telepathically. "I freaked out. I couldn't be calmed down by the medics, and so they sent me to Zendo," he told me. "One of their sitters gave me a hug and helped me process my experience for six or eight hours. Without her, I could have gone to jail that night. The next day, I went back to the tent and volunteered to help." (His friend, by the way, had a great time at the festival until she got a phone call from security about Tobey.)

People have bad trips at home, too, and for them, the Fireside Project offers psychedelic peer support over the phone. What happens, asked the executive director, Josh White, rhetorically, when "somebody watches a Netflix show about mushrooms, eats 5 grams, and then freaks out? People may be terrified of being alone and afraid of dying, not realizing they will probably be okay." And this can happen because perhaps even the personality disorders some people trip to relieve can increase the risk of a bad trip. In that case, they can call or text 62-FIRESIDE, and a volunteer—not a guide or therapist—provides "human-to-human support," sometimes for hours on end. Calls can be in Spanish or one of several other languages, with volunteers who might share identities, like BIPOC and LGBTQ individuals and veterans.

The Fireside volunteers don't have a specific protocol, they're "just tuned in to the emotional experience of people on the other end of the line," and try to deescalate whatever is going on. So, folks flipping out in a large noisy party might be encouraged to find a quiet, calm place for a while. "People think, *All I have to do is eat mushrooms and be cured of depression*," said White. "No, you may have ego dissolution and feel you have died. You may not be prepared to process that." And indeed, many of Fireside's callers are no longer tripping but are having a very hard time processing their trip—including people who have been abused by their facilitators and then gaslit—and so they call for help with integrating their experience.

Folks who have endured a bad trip, even if they don't have serious residual issues, often benefit from talking about their experience. It can help the tripper process the whats and whys of their trip, to hopefully wrap up lingering worries, to grasp whatever learnings might be had. And it is a process that sometimes continues for the rest of their lives. After Tobey told me the story about losing his friend at the music festival, I said, "Wow, it's like you were projecting your fear of not being safe onto your friend." Tobey was silent for a minute. "You know," he said, "that's the first time I've thought of that. It's been ten years since that bad trip, and I am still integrating."

4

Keeping Trips on Course

The fear of enduring a bad trip makes many people nervous about taking psychedelics. But the risk can be mitigated by removing the obvious triggers mentioned earlier and by observing certain precautions—in effect, setting the stage of a trip for the best possible outcomes.

Here's how I think I have skirted the bad trip adventure. I don't have any known preexisting conditions that could be triggered by psilocybin or interact with it in a discomfiting way. I've never gone into a session with a hangover or a cold, or when I was sleep-deprived. I'm careful about where and with whom I trip—though one time, the person who I thought was to be my shaman guide obviously didn't have the same idea and took a humungous dose, and it turned out he was the one who needed care. I also pay attention to dose. Since there is a correlation between anxiety (a potential bad trip trigger) and larger dose sizes, I've opted for fewer fireworks. Less of a trip maybe, but also less anxiety.

All this may sound like good sense, but the truth is there's no guarantee you won't have a bad trip regardless of how carefully the experience is planned. But risk evaluation can make a great deal of difference. That's because the choices you make can enhance or jeopardize the experience. Indeed, much is affected by **set and setting**.

The term *set and setting*, wrote Ido Hartogsohn of Bar-Ilan University in Israel, is a bit of a catchall phrase for the psychological and environmental factors shaping drug effects, and it is fundamental to psychedelic drug research today. "The set and setting hypothesis," he wrote, "basically holds that the effects of psychedelic drugs are dependent first and foremost upon set (personality, preparation, expectation, and intention of the person having the experience) and setting (the physical, social, and cultural environment in which the experience takes place)."

Ideally, dose is a function of mindset, what one hopes to experience. The setting is about optimizing that experience by limiting exposure to danger and enhancing it with an environment that brings joy. The specifics of set and setting are variable, but the message is consistent: first, trippers need to be responsible for their own minds; and second, they need to ensure the safety of their environment and the reliability of the people around them. How that is accomplished is a matter of understanding what is known, what is unknown, and what is unknowable about psilocybin. There is a wealth of anecdotal recommendations that can help inform those choices, starting with the practices of the indigenous healers who originated the magic mushroom ceremony.

The Origins of Magic Mushroom Use

The first time most North Americans heard about magic mushrooms was in a 1957 *Life* magazine article written by amateur mycologist R. Gordon Wasson, the same year *West Side Story* became a Broadway hit. Wasson's article described a mushroom ceremony conducted by an indigenous shaman in a remote Mexican village, a practice that probably evolved from an ancient Aztec ritual first recorded—in a European language—by sixteenth-century Spanish monks.

According to the monks, the Aztecs called the mushrooms *teonanácatl* (from the Nahuatl language, *teotl* means "divinity," *nanácatl* means "mushroom"), and they revered it for its mystical powers. The monks, however, saw the mushroom ritual as unholy and tried to suppress it. As scholars have pointed out over the years, the church was never going to condone any independent communion between humanity and God. Four hundred years later, a handful of mycologists and ethnobotanists, including Wasson, became curious about the identity of the Aztec mushrooms, which led them to villages in Mexico's Sierra Mazateca, where the ritual had quietly endured.*

Wasson's ceremony was led by the Mazatec shaman, or *sabia*, María Sabina Magdalena García in her village, Huautla de Jiménez.** He and a few others spent the night in a humble thatch and mud house, tripping on locally collected magic mushrooms. María Sabina and her daughter, who also consumed the mushrooms, sang and prayed and chanted in Spanish and Mazatec for the duration of the trip, while in the dark Wasson watched kaleidoscope-like patterns organize into palaces, arcades and gardens, mountains, river estuaries, a sea. In his article, Wasson explained the role the ceremony, or *velada*, played in village culture and how the mushrooms were used to provide insight and advice to those in need.

The *velada* is the product of generations of practice. But

---

* Several groups practiced the mushroom ritual: the Nahua in the states of Mexico, Morelos, and Puebla; the Matlatzinca in the state of Mexico; the Totonac in Veracruz; and the Mazatec, Mixe, Zapotec, and Chatino in Oaxaca.

** María Sabina preferred to be called a *sabia*, or wisewoman, over a *curandera*, or healer, according to Álvaro Estrada in *Vida de María Sabina: La Sabia de los Hongos* (1977).

lacking that cultural context, for better or worse, nonindigenous trippers are obliged to shape their own ceremonies. In many cases, like clinical trials, retreats, and sometimes private sessions, the protocols of these trips are informed by the way the Mazatec have tripped for a millennium.

In contemporary therapeutic trips, patients are encouraged to set an intention, which is like the indigenous practice of petitioning the mushrooms for answers and insights into whatever ails them. Trippers cover their eyes, which keeps their attention focused inward, analogous to the Mazatec practice of tripping in a dark house at night, and often listen to music on headphones. A doctor or therapist stays with the patient, ensuring their safety. In a *velada*, the *curandera* conducts the ceremony, sometimes tripping as well, containing the tripper's experience with prayer and song. A *velada* is a spiritual experience: after the trip, the participants might receive instructions to burn a candle for a saint in church. After a therapeutic trip, the patient engages in therapy.

Wasson's article laid out many of the ideas that animate mushroom enthusiasts today: that the mushrooms hold the key to extrasensory perception, that they may be the secret behind the ancient mysteries, a "detonator of new ideas" for early man (an idea developed further by the late ethnobotanist Terence McKenna in his book *Food of the Gods*), even planting in us "the very idea of a god." It was read by millions, as was an adjacent piece in *This Week* magazine written by Wasson's wife, Valentina Pavlovna Guercken, a pediatrician, called "I Ate the Sacred Mushroom." This was the first time most people had heard about magic mushrooms, and they captured the imagination of the American public. Even though Wasson wrote about how the mushrooms were never eaten frivolously or "for excitement," his article inspired droves of American hippies and altered-states seekers to visit María Sabina and partake of her mushrooms.

They were looking for enlightenment or to get high or both, but unfortunately many of those seekers disrespected Huautla de Jiménez's conservative rural culture. Their antics, like making love in the maize fields and wandering around day-tripping, disrupted the village's social dynamics and undermined the sanctity of the mushroom custom. The hippies were eventually kicked out by the army and banned from returning, a prohibition that remained in force until 1976. But María Sabina, who was blamed for all the trouble, suffered: her house was burned down, her son murdered, and her reputation tattered by the "stigma of being involved in sell-out tourism," wrote the poet and scholar Heriberto Yépez. María Sabina's own reasons for sharing the mushroom ceremony with Wasson and subsequent foreigners were complex, suggested the Mazatec historian Osiris García Cerqueda. "She had to fulfill the work that the sacred world assigned her as a means by which the mushrooms could speak and heal people." María Sabina died impoverished and ostracized in 1985.

Despite the Mexican government's disapproval, Americans never stopped traveling to Mexico to participate in mushroom ceremonies, though over the decades their motives seemed to have shifted from the pursuit of a new high to an interest in the *velada* as cultural artifact. I asked Pam Kray, the director of the 2002 documentary *Mushroom Seekers*, about her visit to Huautla in 1999. I met with her at the Rosendale Theatre in the town of Rosendale, New York, once a booming center for natural hydraulic cement mining, now the kind of sleepy place where butterflies float down Main Street. Kray is a tiny woman with white hair and tattoos, shyly articulate but also bold. She participated in a mushroom ceremony presided over by a woman who claimed to be the niece of María Sabina. The *curandera* Augustina "spoke better Mazatec than Spanish," said Kray. Augustina's grade school–age daughters helped their mother during the ceremony. They claimed that the Rolling Stones and the Beatles had tripped in the very same wood-and-mud hut. Clearly, lots of people had tripped there before: the walls were covered in graffiti, "but if John Lennon was there, he didn't sign his name." Kray's experience was like Wasson's: Augustina set up an altar with pictures of Catholic saints, and she prayed and chanted for most of the trip, until the girls made her go to bed. Kray tripped all night in the hut, the latter part alone. "That reset my clock for two or three months," she told me.

Ongoing Cultural Appropriation

Magic mushrooms had long been in use by indigenous people when they were discovered by science. But today, the Mazatecs' traditional use of *Psilocybe* mushrooms has been largely co-opted by modern consumers, and it's looking like some people stand to make a lot of money from that appropriation. The authors of the paper "Ethical Concerns About Psilocybin Intellectual Property" wrote: "From an indigenous perspective, psilocybin research and drug development tell a story of extraction, cultural appropriation, bioprospecting, and colonization."

The indigenous discoverers of psilocybin tested its efficacy on their own bodies. As Terence McKenna pointed out in his book *Food of the Gods*, "If a plant has been used for thousands of years, one can also be fairly confident that it does not cause tumors or miscarriages." That confidence, which nonindigenous people and the psychedelic marketplace enjoy, is based on generations of indigenous experimentation and implementation. But what have the indigenous people gotten? Not much.

The 2007 United Nations Declaration on the Rights of Indigenous Peoples gives the Mazatec a right to their cultural traditions and customs, the right to protect and preserve those practices, to give or deny consent for others to use them, and to benefit from the development of related medicines. But as of 2021, none of the many registered psilocybin-related patents include an agreement recognizing the people who discovered and developed the mushroom practice.

Indigenous people have articulated what needs to change in the paper "Ethical Principles of Traditional Indigenous Medicine to Guide Western Psychedelic Research and Practice," published in the *Lancet* in 2023. The authors are an indigenous-led group of global practitioners, activists, and scholars. Together, they have developed ethical guidelines for the use of traditional indigenous medicines in Western psychedelic research and application. Their paper proposes "accountability for perpetuation of harmful practices and a responsibility for inclusive and respectful practice," "formal efforts to establish Indigenous-led intellectual foundations in Western Psychedelic Science, therapy, and curricula," "benefits for any use of Indigenous medicine and practices are shared with Indigenous source communities," and "therapies based on Indigenous wisdom reorient attitudes towards better relationships with human, other-than-human, and Mother Earth," the last reflecting the original philosophical context of psychedelic use, among other guidelines.

A variety of advocacy organizations are trying to actuate

these principles with indigenous partners. The Chacruna Institute's Indigenous Reciprocity Initiative supports twenty grassroots organizations in the Sierra Mazateca. Another, the Indigenous Medicine Conservation Fund, supports conservation of and education about traditional medicines, as well as harm reduction from social and ecological exploitation. And it is becoming more common for individuals and organizations to make a point of recognizing the lineage of their practice, a modest but symbolically important gesture.

Over the decades, the Mazatec ceremony has been translated, morphed, and sculpted to suit a variety of nonindigenous needs, including healing from emotional and in some cases physical distress, spiritual enlightenment, mental enhancement, wellness (or "mental flossing"), and entertainment. Today, a psychedelic experience can be had by most anyone who wants, under any circumstances they want. Which means the burden is on the tripper to understand the risks involved.

Drug Interactions

As mentioned earlier, the general consensus, both professionally and anecdotally, is if you have a diagnosis of schizophrenia or bipolar disorder, then mushrooms may increase your symptoms, sometimes enough to require hospitalization. So, mushrooms are probably not for you. If you have either of these disorders in your immediate family, mushrooms are probably not for you. What's more (and this is an example of citizen psychopharmacology at work), a comment on Reddit from many years ago recommended that if you were diagnosed as a child with conduct disorder, a potential precursor to schizophrenia, then mushrooms are probably not for you. That warning has not come up elsewhere in my research.

Magic mushrooms, and classic psychedelics more generally, may interact with a variety of medications, though the full scope of potential interactions and their side effects is unknown. You can hire a private consultant, like Spiritpharmacist.com, to do a risk evaluation and check the psychedelic's interactions with whatever medications or supplements you take, but there's no scientific consensus on most drug interactions with psilocybin. Nor is the relationship between dose size and a drug interaction clear. (Though it stands to reason that the larger the dose, the more impactful the interaction might be.) And that's frustrating. I attended a Zoom meeting about microdosing organized by God Dose Quest, a group with little internet presence (by design or not). All the spots that could be accommodated by their free Zoom account were filled, leaving over seven hundred annoyed people who couldn't get into the meeting. The profiles I could see were of mixed age, gender, and race (though it seemed to be predominantly white women), the hosts young, diverse, and enthusiastic, and maybe in over their heads. The meeting opened with some sales pitches for services like microdose consulting and products like liquid culture for *Psilocybe* cultivation, but the chat took over, and the questions came hard and fast (and remained unanswered): "What are the interactions with prescription drugs?" "Can you take psilocybin if you have a concussion?"

Part of any tripper's risk assessment is recognizing that there are lots of unanswered questions, because what is known about interactions is very limited. Most antidepressants affect serotonin systems, and because psilocybin affects serotonin receptors, too, some antidepressants taken in combination with magic mushrooms may cause an overdose called *serotonin syndrome* or *serotonin toxicity*. Serotonin syndrome's symptoms are on a spectrum ranging from benign to lethal, from restlessness and sweating to delirium and kidney failure. Serotonergic psychotropics like selective serotonin reuptake inhibitors (SSRIs) and selective norepinephrine reuptake inhibitors (SNRIs) are considered to present a low risk of serotonin toxicity when combined with psilocybin, but SSRIs may interfere with the ability of psilocin to attach to serotonin receptors in our brains, so it could be your trip never transpires, which some people have reported is both frustrating and upsetting. My friend Lili calls it "blue balls of the mind."

Interactions with monoamine oxidase inhibitors (MAOIs) can cause severe serotonin syndrome, as can tricyclic antidepressants and tetracyclic antidepressants (TCAs and TeCAs). It is unknown how serotonin modulators and stimulators (SMSs) and serotonin antagonists and reuptake inhibitors (SARIs) interact, but possibly they can cause serotonin syndrome, too. It's unknown how norepinephrine reuptake inhibitors (NRIs) and norepinephrine-dopamine reuptake inhibitors (NDRIs) interact. Lithium combined with psychedelics is dangerous and may cause seizures. The irony here is that often the folks who want to try magic mushrooms to treat disorders like depression have tried or currently take some of these medications. I have heard many stories about folks who went off their medications in order to try psilocybin, but the therapists I've spoken with were adamant about proceeding cautiously and talking with your doctors if that is what you are thinking of doing.

Some drugs block the action of psilocybin, like triptans and ergotamines (which people take for headaches). So do steroids, calcium channel blockers like verapamil, and antiseizure medications. There are no known negative reactions to nonsteroidal anti-inflammatory drugs (NSAIDs), antacids, asthma meds, insulin, coffee, B-complex and multivitamins, and Lipitor. The efficacy of the anticoagulant warfarin may be altered by psychedelics. In general, interactions between antibiotics and psychotropic drugs are thought to be benign, though taking them together could be tough on your liver. Pregnant persons should not take psychedelics, as tryptamine, which is what psilocybin is, may cause miscarriage. Whether it is okay to take mushrooms, including microdosing, while breastfeeding hasn't been studied. Anecdotally, trippers have reported that doing mushrooms on your period can make you crampier and more emotional and amplify how awful you feel. One woman on the Reddit page r/shrooms wrote, "If a day on your period is a fuck-you-all and fuck-the-world kinda day, then [tripping] is a no no." Age doesn't seem to matter, though older people may experience less challenging trips, according to a study, which suggests they'd be less prone to a freak-out.

It's probably a good idea to avoid alcohol while tripping. Folks in chat groups often note they have a drink early in a trip to curb anxiety or to suppress the effects of the psilocybin, but others report how easy it is to lose track and drink too much, leading to stomach upset and potentially dangerous behavior. And while many people use cannabis in conjunction with psilocybin, that also can produce unpredictable results. THC, the active ingredient in weed, binds with the same receptors in the brain as psilocybin, potentially intensifying the effects of the mushroom. But what about the drug's interactions with other conditions, like heart murmurs, head injuries, diabetes? That's yet to be determined.

Unusual and Everyday Side Effects

Addiction isn't really considered an issue when it comes to periodic use of magic mushrooms, and of all the recreational drugs, they are the least likely to be abused or cause death. Flashbacks are rare, too, but scrolling through the mushroom forums does reveal scattered cases of HPPD, or hallucinogen-persisting perception disorder, which is like flashbacks. Users mainly describe persistent visuals like halos, color flashes, and snow "like TV static," and auditory hallucinations that occur after the trip has ended. These effects can come and go, sometimes for a short period of time, but for some, HPPD can reoccur for months, even years, and it can start after just one trip. "I took the drug to see the world different[ly] for a few hours," said one HPPD sufferer on Reddit, "and I guess I overachieved." Some people who experience mild cases of HPPD don't necessarily find it unpleasant, but others find it debilitating. Combining weed or cannabis edibles with psilocybin may bring on HPPD, and the risk could be higher if you replace the psilocybin with LSD.

Unfortunately, even if a doctor can diagnose this rare condition, there is no established protocol for treating it. Antipsychotic medications could make the symptoms worse. For some people, the condition fades over time. "I've had HPPD for 2.5 years and slowly recovered to a point where I'm completely recovered!" announced one person on the Reddit page r/HPPD. "Do you think I could smoke some DMT or would that be too risky?" For others, certain states of mind, like stress or lack of sleep, can bring back the effects. And some just learn to live with it, joining the ranks of those who go through daily life coping with disorders like tinnitus or eye floaters.

A more typical negative reaction is nausea. On every adult tripping excursion I've attended over the years, at least one person in ten spent a great deal of time retching in the woods. Many people have suggested the culprit is chitin, a largely undigestible component of fungal cell walls, but I think it is unlikely chitin is to blame. Chitin doesn't break down under cooking temperatures, yet a mushroom risotto doesn't make substantial numbers of diners feel ill the way magic mushrooms do. Rather, the psilocybin itself seems to be the culprit. Even in clinical settings where synthetic psilocybin is used—so no chitin present—patients list a queasy feeling among their symptoms. There are a lot of serotonin receptors in the GI tract, receptors that psilocybin affects. "I wouldn't be surprised if the nausea was due to the psilocin directly," said the mycologist Bryn Dentinger of the University of Utah. He thinks when an animal like a slug eats the mushroom, perhaps the molecule functions as an emetic or laxative to ensure the mushroom spores are expelled before being destroyed by the harsh conditions of an animal's gut.

To mitigate nausea, most folks just avoid eating much on a trip day. A Mazatec-led ceremony I participated in required a full day of fasting before an evening trip, but I was so hungry that it dominated the early hours of my experience—my empty stomach contributing to the setting—and led me to think a lot about how the ugly feeling of hunger is all too familiar to many people in the world. On the other hand, a full belly might slow the absorption of the drug, leading to a longer period before the effects kick in. Christopher Hobbs says it can take as long as two hours to metabolize the drug if you've consumed a fatty meal, like osso buco, beforehand.

Equally common is onset anxiety, that "what have I gotten myself into?" feeling. "It's always difficult letting go of this world," pointed out a friend. I would agree. But it is important to nip anxious feelings in the bud, because should they progress and intensify, they can veer you into a bad trip. Trippers can mitigate those first twinges of anxiety by tripping in a safe place with safe people.

Safe Faces, Safe Spaces

The larger the dose, the more vulnerable a tripper will be. And the more vulnerable the tripper, the more important it is to secure a safe place and trustworthy people. Microdosing, as it is typically understood, shouldn't require special accommodations. Many folks who take a low dose, or "museum dose," do so in public spaces and events, but they can get in trouble if they indulge in behaviors where impaired judgment could harm themselves or others, like driving. Based on numerous trip reports, those who take therapeutic- and heroic-size doses are better off when they secure a safe space where they can stay put for the duration of their trip and ensure someone is around who knows what they are up to.

Hopefully, that will be someone trustworthy, because the company one keeps can affect the outcome of the trip, though even the weirdest situations can work out okay. In the film *Know Your Mushrooms* (2008), the late Gary Lincoff told a story about taking magic mushrooms with some folks he didn't know and slowly realizing they had a lot of drugs and guns in the house. He ended up having a memorable trip, but he did get anxious when he hallucinated that his host turned into a wolf.

The people present during a trip contribute to the setting. They can make a setting feel secure or scary. And that can affect outcomes, because a trip can take some people to tough places emotionally, and tripping among people who feel unsafe can lead to anxiety. But there is little consensus on who makes an ideal tripping companion. The mycologist Paul Stamets, in his book *Psilocybin Mushrooms of the World* (in the chapter "Good Tips for Great Trips"), recommends tripping with someone you love. But Timothy Leary, who was married six times, recommended against tripping with a spouse. I once tripped with my husband and pretty much reverted to a sober state because I spent the whole time worrying about how he was doing. A friend of mine said it is important to trip with someone you can cry or throw up in front of, but whom you don't want to sleep with. There doesn't seem to be any rule as to who makes the best kind of trip sitter, as long as they make the tripper feel safe and keep them from doing something harmful.

Trippers are highly influenced by their environment, and it can affect their response to the psychedelic. I've mentioned the city-versus-nature paradigm. Cities can be overstimulating ("Times square on LSD—sounds great, right? It's the epitome of sensory overload," warned one psychedelic commenter) and can make one fretful and lead to misadventures ("Crossing busy streets is pretty fucking difficult," said another). But for those who do choose a city setting, trippers advise not to do anything *else* illegal, like jaywalking, and to bring food and water, because "the last thing you want to do is wander into a restaurant for something quick and find yourself stuck."

Using magic mushrooms in nature is not without its

challenges either. As Rachel Clark of Lucid News points out, it is all about dosage. "The higher your dose of any psychedelic," she wrote, "the less capable you will be of problem solving." Because anything can happen in the mountains. The online forums are full of good advice, like "Don't take your first trip in a place where the only help comes in the form of a search and rescue team with a helicopter," and "Don't go walking around at night. Spider webs. They're everywhere. Or at least you will think so after walking into one." And for those who plan to trip while camping, "get your campsite dialed in before lift-off. Setting up a tent while tripping is . . . bewildering."

Determining an appropriate setting is ultimately about what feels right. For example, is the tripper a day person or a night person? Night trips are the norm in the Mazatec ceremony, and most clinical trials and some retreats call for day trips with eye masks, but many people prefer day trips with their eyes wide open. The Mazatec ceremony utilizes prayer, smoke, and sacred objects to create a setting, but private groups that come together to trip often create their own ceremonial space.

The Art and Science of Conducive Atmospheres

One way you can determine if a legal psychedelic retreat is a good fit for you is their approach to setting. Some feature a luxurious spa-like ambience; others emphasize the natural world, with access to gardens and groves. At research hospitals and universities where psilocybin is studied to treat disorders like addiction or depression, the settings are intentionally controlled to eliminate one potential variable.

Music contributes to the setting, too, from the singing and chanting *curandera* in a Mazatec ceremony to performances by psychedelic jam bands like the Grateful Dead and Phish or concerts of ambient and electronic sounds that ripple through reclining audiences. Tripping music is something of a cottage industry, consisting of mixes from the likes of Cosmic Gate and albums like *Music for Mushrooms*. Some retreats and guides offer live music performances in person or on Zoom that you can sync your trip to. Patients in many clinical trials are offered earphones with a seven-hour playlist. The one at Johns Hopkins was masterminded by William Richards, whose involvement in psychedelic research dates to 1963. It includes "Ascent," "Peak," "Post-Peak," and "Welcome Back to Earth" music, which wraps with Louis Armstrong singing "What a Wonderful World." You can hear it on Spotify. "Sensitivity to the therapeutic potential of carefully selected music," wrote Dr. Richards in *Sacred Knowledge*, "may be an important factor in enhancing psychological safety."

Music can also create negative settings. A friend of mine told me about a high-dose trip where the guides played ominous music and were growling. "There could be a utility to that," he said, "but there's an ethical expectation to make sure it's okay." At the peak of one nighttime session I attended, someone started playing a sound bowl. The ringing was so intense, I sat up with my fingers in my ears, and I began to worry that the psilocybin had caused some kind of brain injury where I couldn't stop hearing the ringing and I never would stop hearing it. When the ringing finally stopped, I wept in gratitude.

The negative effects of music have also been noted by scientists. In his study on the efficacy of psilocybin to treat cocaine addictions, Peter Hendricks found the mood of participants changed immediately depending on the music they were exposed to. In a presentation at the Telluride Mushroom Festival in 2022, he told a story about a man who was not responding well to the psychedelic. Turns out, he didn't like the classical music on his headphones. When the researchers changed the music to gospel, however, the participant's response to the drug improved dramatically.*

Location, Location, Location

A trip lasts awhile, and since it can be difficult for some people to change locations midtrip, it's important to secure a setting for the duration. That includes checking the weather forecast. One group of trippers I know thought they had selected a beautiful setting on a mountainside in Colorado until they were drenched in cold rain and getting off the mountain required jumping into the cab of a continuous gondola, a timing challenge one tripper found baffling. A night setting where it is impossible to hang out for the duration of a trip or sleep afterward might mean driving while still in a vulnerable state. Dr. Fadiman's advice is to set aside three days for a trip: one to prepare, one to trip, and one to integrate the experience. If this sounds like a lot of commitment, it is. That's the reason why it's called a *trip*.

* There is also a history of visual co-therapy. Therapists of the 1960s like James Fadiman recommended having a rose on hand for trippers to sink into. Today, the Pacific Neuroscience Institute has studied the effects of nature-themed immersive video (by *Fantastic Fungi* filmmaker Louie Schwartzberg) on trippers' experiences. They found the images helped calm and ease the participants in the early stages of their trips.

Consuming the Mushroom

Unless they are participating in a research study in a clinical setting, trippers are likely to consume psilocybin in the form of actual mushrooms, either fresh or dried, whole or powdered. (Clinical trials use synthetic psilocybin, a substance that is tightly regulated and hard to come by.) I've encountered psychedelic guides who make a point of serving only the caps, believing they are more potent, but the environmental chemist Michael Beug, who did seminal work on the psychoactive properties of *Psilocybe* species at Evergreen State College in Washington, told me his findings showed the tryptamines (psilocybin, psilocin, and the other psychoactives present) are evenly distributed throughout the mushroom.

Microdoses are almost always powdered and either put in a capsule and swallowed or added to a drink or food, like a spoonful of yogurt. For larger doses, many people just chew up the dried whole mushroom. The caps are generally easy to rehydrate with saliva and nibble into small enough bits to swallow, but more than once, I have ended up leaving my stems behind, like the tough ends of asparagus on a plate.

Some people eat the mushrooms fresh, though this option can carry particular inconveniences. Wild species can be wormy, and even cultivated specimens may contain fly larvae. (To deal with worms, according to Dr. K. Mandrake, one of the authors of *The Psilocybin Chef Cookbook*,* leave the mushrooms

* I review cookbooks periodically and always test a selection of recipes to determine if they work and taste good. Usually, I can test a few recipes a day, but in the case of *The Psilocybin Chef Cookbook*, I had to set aside a whole day to test one recipe. I tested a Shroomshuka, a tomatoey sauce with eggs poached in it, with half a gram of dried

somewhere warm—but out of direct sunlight, which degrades the psilocybin—"and the worms will make their way out.") Additionally, the ratio of dried psilocybin to fresh mushrooms is about the same as mushrooms generally: one to ten. A high dose of fresh mushrooms would be about 30 grams. All those raw mushrooms can be hard on the stomach, which is why many people cook them first.

Fresh mushrooms can be added to any dish that calls for mushrooms, but *P. cubensis* really aren't that tasty. They taste like raw flour. Kinda gross. Other people have described the taste as metallic, earthy, bitter, and like raw seeds. So, clearly, taste is personal. For example, the author Andy Letcher described liberty caps as "greasy." "He is trippin'!" exclaimed a friend when I mentioned Letcher's opinion. "*Cubensis* taste so much worse than liberty caps!" To me, dried *Psilocybe* aren't much better. They taste like cardboard with a mild case of foot fungus. Folks get around the basic unpalatableness of dried mushroom by grinding them up and sprinkling the resulting powder into recipes, packing it into gel caps, or steeping it in hot water to make a tea. (Though teas are only "good" for a day, according to Paul Stamets. After that, "ethyl alcohol must be added to prevent fermentation.") While some sources insist that psilocybin degrades when steeped in boiling water, a paper in 2020 found that powdered mushrooms can withstand boiling temperature (212°F) for thirty minutes with little loss of potency. The wellness company Numinus has developed a mushroom tea bag—for clinical research right now, but one can imagine someday opening your sister-in-law's kitchen cabinet to find a box of psilocybin tea next to

psilocybin sprinkled in. It's the kind of dish somebody living in college housing might whip up: one pot, a few ingredients, and the ability to get high.

the Sleepytime. Obviously, injecting the tea intravenously, as one fellow did, is a terrible idea. It led to acute liver injury and failure of his kidneys and lungs.

Lots of people make a lemon tek, where the juice from one lemon is steeped with a gram of powdered mushrooms for twenty to thirty minutes, then strained. They claim the acid from the lemon converts the psilocybin into psilocin, leading to a faster delivery of the effects, maybe even intensifying them. Chocolate may intensify effects, too. The Aztec consumed cacao with the mushroom, and when I attended a ceremony conducted by a Mazatec *curandera*, we were given a few cacao nuts with our mushrooms. Cacao is a potential MAOI, which might intensify and elongate the trip because it slows down the natural disintegration of the psilocin molecule.

The proliferation of mushroom-spiked chocolates might be a direct reflection of entrepreneurs capitalizing on the Aztec tradition. Or maybe it's just chocolate. Many people make their own confections (there are a few recipes in *The Psilocybin Chef*), but there's no regulation on commercial products, so unless you send it to a lab for analysis, it's impossible to know for sure there even are *Psilocybe* mushrooms in the chocolate bar you bought. What's more, the dosing recommendations on the package just might not apply to you.

Putting It All Together

Psilocybin can be a tool to unlock the mind, given the tripper considers her dose and intention, and successfully mitigates her risks, as my friend Vivian, who took a heroic dose, knows. She set out to explore parts of her consciousness that had resisted disclosure, and she succeeded, though in ways she didn't expect.

Lots of people put the heroic dose of psilocybin into the framework of the writer and professor Joseph Campbell's hero's journey. Campbell believed that all humans share a set of "elementary ideas." His book *The Hero with a Thousand Faces* (1949), a work of comparative mythology, illustrated what he called the hero's journey monomyth—the idea that all hero myths are variations of one greater narrative. Campbell broke down the heroic journey into stages: The hero sets out from the ordinary world when she or he is called to adventure. A mentor helps her cross a guarded threshold leading to the supernatural world where none of the usual rules apply. She undergoes a road of trials where her mettle is tested—sometimes benefiting from allies—and finally a most challenging ordeal. She reaps a boon, like a blessing or advantage, because of the ordeal and then heads back to the ordinary world a changed person, one furnished with the means to improve her world. Think Ulysses, Buddha, Black Panther, Luke Skywalker. Vivian.

She is a retired lawyer and hobby farmer, affectionate and dignified, a rosy-cheeked lover of cooking and foraging. She chewed through 6 grams of dried *P. cubensis* in a meadow in the Rocky Mountains. Vivian is not inexperienced; she trips every few years. Nor was 6 grams the largest dose she had ever taken. She was with friends she knew well, and she was totally prepared: she had a raincoat and water and snacks—that's Vivian. When she brings a dish to a dinner party, all the components are ready, the garnishes and platters and tongs. I always considered her the kind of person who thought of everything.

But her default physical state is clenched up, with shallow breathing, "even when I am milking the goats," she told me, as if that came to her as a surprise. Vivian brought a very specific intention to her trip. “My intention was to find out why I am always clenching up.”

When she reached the stage of her trip where, for some people, deep memories flood into consciousness, she observed herself in a scene from when she was four. She rode her tricycle to the house of a neighbor whom the local kids called the Candyman to see his dog. He kidnapped and raped her, and the police found her in his house stuffed into a blanket chest. He was arrested and tried but the jury did not convict, nor did she—a child—testify. “I never wanted to examine it,” she told me, “but during the trip, I saw how innocent and trusting I was. No one had ever hurt me before. I toughened up to get through life. I said, ‘I am okay; this didn’t impact me,’ but I still held on to it.” She realized she has been in flight mode for fifty years, clenched up, ready to bolt.

In the framework of Campbell’s heroic journey, that insight was the boon she reaped from the ordeal of reconsidering her abduction. With it, she headed back to ordinary reality furnished with the means to improve her world. After the trip, she decided to examine the crime and look at the court records she’d never read. There would be no more dismissing what had happened, and this, she felt, might help put the past behind her.

Vivian’s experience exemplifies the promise of psilocybin, but it also reminds us of the key role that mindful preparation can play.

PART II

In Search of the Mushroom

5

Foraging

Hunting magic mushrooms in nature or cultivating them at home are the most private ways to acquire magic mushrooms. It's just you and the mushrooms, in the woods or field, or at home with a cultivation rig. I love mushroom hunting, and so that is my preferred way. Generally, I am what's called a *pot hunter*, someone who hunts for culinary mushrooms. It's ultimately acquisitional: pot hunters aren't very excited by species we can't eat. I like to hunt magic mushrooms, too, because finding them scratches that same acquisitional itch.

Magic mushroom hunters are experts at observing habitat and environmental conditions, and anticipating where and when the different species might grow, even what conditions may change their payload of psychoactive ingredients. Indeed, hunters are responsible for much of the knowledge we have about these species today.

I've met many people who eat psychedelic mushrooms (or want to) but really have no clue about the nature of the organism, where and how it grows, why psilocybin and other alkaloids are present, and how those alkaloids vary among species. But anyone interested in acquiring magic mushrooms by foraging must start with a little mycology. From there, it's just a matter of finding your teachers.

Magic Mushroom Mycologists

I attend mushroom hunting forays throughout the year, though usually not forays with robust magic mushroom collection. But in the fall of 2022, I attended one in the quaint seaside town of Ocean Park, on Washington's Long Beach Peninsula. It was cranberry season, and along the main road, kids were selling the bright red berries stuffed into two-gallon baggies like cold, nubbly pillows. Long Beach is a skinny peninsula bounded by the grand Pacific Ocean and the rich estuary waters of the Willapa Bay, a strip of sandy land covered in dune and grass and cranberry bogs, verdant cow pastures, pony corrals, and conifer forest. I was there for a foray that celebrated every kind of mushroom—cinnamon-scented matsutake and chubby porcini, of course, but also modest little psychedelic mushrooms whose charm is all about what they hold inside.

The foray was based in a rented house that was blessedly close to the ocean and the scrubby woods that abut the beach. I stayed at a hotel nearby, which was a good idea, because over the course of the weekend, a tumbled pile of hiking boots by the front door grew hourly as more and more people crashed at the house, which was alcohol-free but tripper-friendly. Along with twenty or so other people, we spent the days hunting mushrooms in the dune grass, the woods, and the pastures, and as the weekend unfolded, a specimen table became heavier and heavier with mushrooms set on paper plates for identification. There were numerous psychedelic species on the table, all in the *Psilocybe* genus.* That's not

* So far, seven genera besides *Psilocybe* have been determined to include psilocybin-containing mushrooms: *Conocybe*, *Pholiotina*, *Galerina*, *Gymnopilus*, *Inocybe*, *Panaeolus*, and *Pluteus*, according to "An Overview on the Taxonomy, Phylogenetics and

surprising; there are maybe 200 species of psychedelic mushrooms overall, but most them, around 165, are in the genus *Psilocybe*.

The name *Psilocybe* comes from the Greek, meaning "bareheaded," and refers to the fact that many species have a detachable pellicle, like a peelable skin on its cap. Almost all species of *Psilocybe* contain psilocybin, psilocin, and other psychoactive compounds, but they vary in size and shape, habitat, and potency. These species, and a few other psychoactive species from other genera, are often lumped together and called *magic mushrooms*—a term coined by the editors of R. Gordon Wasson's 1957 story in *Life* magazine. The US Drug Enforcement Administration lists all kinds of slang names I've never heard used, like *mushies*, *psilly billy*, and *stemmies*, and some I have, like *shrooms*. Maybe undercover police officers use those slang terms, but most people who are serious about magic mushrooms use the Latin binomial. It's the gold standard because the genus and species tell you exactly what mushroom you have, and from there, you can determine its potency. In some cases, though, common names will refer to species, like liberty caps, which are *Psilocybe semilanceata* (a strong psychoactive mushroom relative to others), and blue ringers, which are *Psilocybe stuntzii* (a weak psychoactive). But if you were to examine a "shroom" under the microscope, it would likely be *Psilocybe cubensis* (a moderate psychoactive) because that's the most widely cultivated species.*

Ecology of the Psychedelic Genera *Psilocybe*, *Panaeolus*, *Pluteus* and *Gymnopilus*." These mushrooms are mainly consumed by people who collect them in the wild. Also, the fungus *Massospora cicadina* is a parasite of cicadas, deteriorating the insect's abdomen and releasing, among other chemicals, psilocybin. No one, to my knowledge, eats the infected cicadas.

* I pronounce the group *si-low-sigh-bee*, but there is no definitive pronunciation, as Latin words have been transliterated into many languages with different pronunciations. I use the pronunciations I do because that's the way my teachers pronounced

In the 1950s, mycologists like Roger Heim and Gastón Guzmán began identifying species in the *Psilocybe* genus. They are found all over the world, but most species live in the neotropics, the tropical areas of North, Central, and South America, especially Mexico. The Aztec magic mushroom, the *teonanácatl,* likely refers to a group of psychoactive species, though according to the mycologist Laura Guzmán Dávalos, *Psilocybe mexicana* was the one most commonly used in Mazatec ceremonies and the mushroom that Wasson consumed. But as María Sabina said in the 1978 documentary *María Sabina: Spirit Woman* by Nicolás Echevarría, "There are different kinds of children. Those that grow on sugarcane, those that grow on trees covered in mold, and those that grow from the damp earth."

Where Do *Psilocybe* Mushrooms Grow?

Psilocybe are saprobic fungi; they eat dead and dying plants, and as the nutrients are consumed, the fungus makes mushrooms with spores that move the fungus's genetic material to a new rotting log or pile of dead leaves. In the US, *Psilocybe* species thrive in the coastal and southeastern states, preferring floodplains, woodlands, moist fields, grass, dung, and man-made habitats like landscape wood chips.

I've heard collectors of *Psilocybe* say the different rotting habits of particular species affect how the trip feels when the mushrooms are consumed. For example, *P. azurescens* and *P. cyanescens,* which break down wood and dead grass, have a strong "body

---

them. Anyway, don't let scientific pronunciations get in your way, because the Latin binomials are useful.

load." In trip-speak, that means you might just need to lie down for the duration. *P. semilanceata* and *P. ovoideocystidiata*, which consume dead wood that is on its way to becoming soil, like mulch and wood chips, and *P. cubensis*, which consumes herbivore manure (manure being mostly decomposed plant material), have a lighter body load or no body load at all. (Note that this is outside the hallucinatory aspect of a trip, and bear in mind that how anyone responds to any amount of mushroom is totally individual.) Finding these mushrooms, and really any mushroom, is a matter of looking in the right place. Mushrooms are the reproductive organs of certain fungi, and since a fungus lives in its food, to find the mushroom, look for the food.

On the Long Beach foray, we found four common *Psilocybe* species. The foray chef, Sebastian Carosi, is a friend, but as he was busy in the kitchen making delicious things like pork chops with porcini gravy, he fobbed me off on a local fellow I'll call Jones. Jones took me and a few others on a beach hunt for *Psilocybe azurescens* (azzies), a delicate and fragile but quite potent psychoactive mushroom that feeds on the roots of dead dune grasses. I figured our group of four would tromp around, as we usually do when hunting mushrooms, but instead we loaded onto electric bikes and, like a small gang, zipped along the curvy asphalt paths that lay like a ribbon over and around the dunes, looking for patches of dying grass. Standing on our foot pedals to get a wider view, we peered along the path edge for signs of azzies, and when we spotted one, we waded into the hip-high grass, the ocean gently roaring in the background. And the smell! Pine needles and salt water and horse manure from the riders we shared the path with. "We'll come back here next year," said Jones, gesturing to where we had trampled the grass, "'cause all of this will be dead," providing future food and habitat for the mushrooms. Azzies grow in troops—if you see one, it is likely you will find five more. "But when they have a substantial flush," Jones began, and then he smiled and shook his head. "Everywhere."

Over the course of the weekend, I made several outings in my rented car with various other mushroomers, eyes peeled for *Psilocybe* habitats like the degraded wood chip piles we found in vacant lots, prime habitat for the potent *Psilocybe cyanescens*. Its common name is *wavy cap* because the cap undulates like a twirling skirt. We looked for lush cow pastures, the grass growing deep green, moist, and puffy, the perfect habitat for little *Psilocybe semilanceata*, or liberty caps. Their nickname comes from the shape of the cap, which looks like the bonnet rouge of the French Revolution.*

Liberty caps are present in at least seventeen countries and, according to the late mycologist Gastón Guzmán, are probably the most popular and abundant of the psychoactive species in northern Europe if not the world, with a fairly consistent psilocybin content no matter where they are collected. They are considered moderately active to extremely potent.** The psilocybin also seems to remain stable in dried mushrooms, even for a very long time; psilocybin has been detected in specimens as old as 115 years.

The fungus feeds on the dead roots of grasses like fescue, sedges, and perennial rye (so you might find it in lawns and

* But there's more. The bonnet rouge, or liberty cap, is derived from the Roman pileus, a cap awarded to freed slaves. The scientific name for all mushroom caps is *pileus*. This is the kind of random association that makes trippers think there is an order to everything in the universe.

** The presence of psychoactive compounds is determined by the chromatographic method, which separates components (like psilocybin and psilocin), allowing their percentages to be measured.

sports fields, too). When we clear woodlands to create pasture for our animals, and those animals enrich the soil with their manure, we create the ideal habitat for liberty caps. "Where we have walked," observed Michael Beug in a conversation, "*Psilocybe* have grown in our footsteps." If you see people in the Pacific Northwest meandering around in grassy pastures with the telltale "Psilocybin stoop," it's likely they are hunting liberty caps. The stoop is so well known by local farmers that when my companions and I hit a particularly promising pasture, they told me to act inconspicuous and walk upright but with eyes cast down, which made us all look stiff and hesitant, like zombies crossing the field. (Ideally, if you want to hunt on someone's property, it's best practice to ask first.) It made finding them more difficult, but my fellow hunters were experts at spotting the mushrooms. As we stumbled about, for every couple of mushrooms they put in their pockets, they popped one into their mouths. Maybe that's why, during a hunt for bioluminescent fungi later that night, most of the foray members wandered off and got lost.

Identifying *Psilocybe*

For a couple of years in the early 1970s, Mike Maki, whom the DEA dubbed "the Pied Piper of mushrooms," and a few friends who had figured out how to identify liberty caps had the magic-mushroom-picking scene in coastal Washington all to themselves. "It was heaven for a minute," he told me. Then, in 1973, Timothy Egan, a student at the University of Washington in Seattle (and a distinguished nonfiction writer today) wrote a piece in the student paper about the "new craze for magic mushroom picking" of wavy caps in the landscaping wood chips scattered around the dorms and beside the walkways. The craze spread beyond campus as hunters realized azzies were growing in the Astoria area and liberty caps in the fields, especially low-lying tidewater cow pastures, said Maki. By 1975, so many liberty caps were collected in the Pacific Northwest that a market emerged, enabling pickers to sell their spoils and farmers to charge for access to their fields. It was around this time that *Psilocybe* identification guides began to appear, and reliable volumes like Gary P. Menser's *Hallucinogenic and Poisonous Mushroom Field Guide* (1977) and Paul Stamets's *Psilocybe Mushrooms & Their Allies* (1982) became available to trip-seeking foragers.

Psilocybe are little brown mushrooms (LBMs for short), and it is important to note that LBMs have deadly look-alikes. While no toxic or lethal *Psilocybe* species are known, several poisonous lookalikes share the same habitat. For example, *Psilocybe stuntzii* is a common Pacific West Coast psychedelic that grows in decaying wood debris in the fall. Unfortunately, so does *Galerina autumnalis*, the deadly galerina. They look similar: small, brown, and gilled, though the *Psilocybe* stains blue when picked and the *Galerina* has a brownish or blackish base. But if you didn't pay attention to the gill color or bother with spore prints—*Psilocybe* have purplish-brown spores, and *Galerina* have rust-colored spores—and ate enough of the wrong mushroom, whether dried or fresh, raw or cooked, you could suffer a kind of poisoning similar to consuming the death cap (*Amanita phalloides*) or the destroying angel (white members of *Amanita* sect. *Phalloideae*), which can lead to liver and kidney failure. In 2023, a potential antidote for amatoxin poisoning, the medical dye indocyanine green, was discovered. Too late, unfortunately, for a teenage girl who, in 1981, died from mistaking a deadly galerina for a magic mushroom.

The blue staining is the most distinctive characteristic of mushrooms that contain psilocybin. When handled, the stalk develops patches of cyan pigment the color of a swimming pool. (It's the shade worn by many nurses and surgeons, and you, too, when you put on those awkward paper gowns. Even hospital rooms may be painted cyan. That's because cyan, among other benefits, is thought to calm and relax patients.) Paul Stamets says, "If a gilled mushroom has purplish brown to black spores, and the flesh bruises bluish, the mushroom in question is very likely a psilocybin-producing species."

The blue staining seems to be the result of oxidizing psilocin. In 2019, the German microbiologist Dirk Hoffmeister and his team discovered that when *P. cubensis* is injured (picking is enough to do it), it can trigger a cascade of enzymatic reactions. One enzyme converts the psilocybin into psilocin, and another enzyme, triggered when exposed to oxygen, accelerates the oxidation of the psilocin, which makes it unstable. The unstable psilocin molecules join to create structures that appear blue to our eyes. The oxidation of psilocin also undermines the mushroom's overall potency as a hallucinogenic. "I doubt, though," said Dr. Hoffmeister in an email correspondence, that "this is relevant in 'real life.'" He means from the mushroom's point of view.

Mushrooms probably produce psilocybin as an antifeedant, though the science is far from established. Bryn Dentinger, whom I caught between *Psilocybe*-hunting expeditions in Guatemala, thinks the psychoactive molecules in magic mushrooms might have evolved to defend the mushroom from terrestrial gastropods like snails and slugs (think critters with a serotonin receptor). "We can date the origin of *Psilocybe* mushrooms to sixty to sixty-five million years ago. After the asteroid [that killed the dinosaurs] hit, *Psilocybe* diversified, as did land gastropods." Around half of all plant species went extinct because sunlight was blocked by dust. But fungi don't need sunlight to grow, so they were around to provide feed for the gastropods. "That's my hypothesis," said Dr. Dentinger, who was planning an experiment. "Actually, right now I'm trying to order some slugs."

While all *Psilocybe* mushrooms stain blue, some blue so little it is virtually impossible to see. Likewise, just because a mushroom stains blue doesn't mean it's psychoactive. *Gyroporus cyanescens*, sometimes called the *cornflower* or *bluing bolete*, is an edible, if alarmingly blue, bolete with no psychoactive properties. (The bluing is the result of oxidation of gyrocyanin, a phenol—not psilocin.) And some psychoactive mushrooms don't stain blue, because they contain neither psilocybin nor psilocin. For example, the psychopharmacological effects of *Amanita muscaria*, the iconic red mushroom with white warts, come from the presence of the GABA analog muscimol. So, there is no definitive blue rule, just blue guidance.

Meeting Mushroom Hunters

Mycology only recently started to attract a wider audience. Of all the forays I've been to over the years, most were populated by retired folks. The crowd at the Long Beach foray, however, was young and tattooed, dressed in variations on the Pacific Northwest uniform of flannel and Carhartts, but underneath they wore their predilections: psychedelic tie-dyed T-shirts featuring pulsating cannabis leaves and jam band logos. This cohort was a study in contrasts: scroungy yet fluent in Latin binomials, partiers with biochemistry degrees who, between hits on their vape pipes, ruminated on the soil-binding capacities of fungal mycelium, the morphology of mushrooms, the haul of chanterelles they had found earlier in the season. Most were foragers, not just of mushrooms but razor clams, sea beans, crabs. I was a bit out of place—an older, sober woman with a notebook—but the other participants were kind and warm and humored me when I asked them to show me the locations where they had already spotted some psychedelic mushrooms.

This turned out to be a patch of the mildly psychoactive *P. stuntzii* (commonly called *blue ringers*) growing in the gravel and sod landscape of a modest home, surrounded by chain-link fence. The guys told me to drive real slow so we could get a look, then park across the street. The day before, they'd waited for the homeowner to drive off, then hopped the fence to inspect the mushrooms. They recommended we do the same this time. But no way was I going to try jumping over that fence, so I knocked on the front door. The woman who answered was about my age, and I explained we were amateur mycologists—and I gestured toward my companions, in their wool caps and saggy jeans, waiting, in a line, across the street, shifting from foot to foot—and wanted to inspect the odd mushrooms growing in her landscaping. She was thrilled to talk about them. They were driving her crazy, she said, and she didn't know how to kill them. I explained they would only come up for another season or two, so she didn't have to throw bleach on them or anything, but if she liked, we would pick them all. Not only was she happy to wave over my companions, but she supplied us with grocery bags to carry the mushrooms away. An hour and a half later, we had picked all the mushrooms, and I met the owner's mother and learned about the grandchildren, a recent stroke, and problems with a nosy neighbor, but also that they had laid sawdust under the gravel. That's probably what the fungus was feeding on, and it suggested the sawdust had been inoculated prior to being laid. I asked where she had purchased it, because at that point, I was picturing huge piles of sawdust erupting with blue ringers. She sent us to the place, Basket Case Greenhouse (I kid you not), but they were closed, and their fence was too tall for any of us to jump over. But the moral of the story is, it pays to ask for permission.

If you want to meet mushroom hunters, join a mycological society. Go to the North American Mycological Association's website, which lists clubs nationwide; there are a few clubs in almost every state. Clubs typically organize guided walks in the woods with an expert. Mushroom identification, I think, is best learned peer to peer, and for good reason. It is hard to use images in a book or an app to identify a mushroom because the vagaries of age and the wear and tear of weather can change a mushroom's appearance just enough to make you feel unsure. (Plus, it could be dangerous. The New York Mycological Society has warned its members about AI-generated identification guides littered with "hallucinations" that can get you sick.) But when you are walking in the woods with someone who knows mushrooms, and they point out a species, you benefit from seeing the spectrum of its morphology—young, old, dehydrated, buggy, sunburned, waterlogged—in its habitat.

Mushroom Hunting Manners

As a forager, you would do well to mind basic mushroom-hunting manners if you want to stay friends with other hunters. It is obnoxious to follow someone and then swoop in on a patch they have discovered. In general, each finder takes home all his spoils unless the hunters have agreed ahead of time to pool their loot at the end of the day. Some hunters are reluctant to share their spots; if they do, that doesn't give you the right to go in there and wipe it out. "I had a *Psilocybe cyanescens* spot in Prospect Park, behind the zoo," said Sigrid Jakob. She's the president of the New York Mycological Society (a volunteer job I held from 2011 to 2016). "I told a young man about it and the next time I went there all the mushrooms had been picked. I found the culprit pushing his baby in a stroller. He ostensibly took his kid to the zoo, but in fact he went and picked all my *Psilocybes*." If a mushroom hunter tells you her spots, be grateful, because for every spot a hunter has found, she has walked miles looking.

Beware of depending on social media for help with mushroom hunting and identification: you will find some good information but also plenty of mycological misunderstandings, like the picking-versus-cutting controversy. Many people are passionate about not disturbing the fungus, but unless you are doing something really damaging like raking up soil, the fungus will be unaffected by picking the mushroom; picking and tromping about may even stimulate more mushroom production, said Dr. Dentinger, as the fungus seems to thrive on a little disturbance. It is probably running throughout a large area anyway, not just where the mushrooms appear. Often, when picking *Psilocybe* species, you must grasp the base of the fragile stalk and yank, which brings up a little clump of fuzzy fungus and soil, like a tiny root ball. Not only is it okay, but you can save those hairy balls and use them to inoculate a new patch (more on that soon).

Spore prints are a useful tool in mushroom identification because they are a way of determining spore color, but not a definitive one: many mushrooms of different species have the same color spore print. For example, *Hypholoma fasciculare* has the same purply colored spores as *Psilocybe* and it is toxic. To make a spore print, place the cap of the mushroom gill side down on a piece of paper, ideally half white, half black, since many spores are white. (You can also use aluminum foil, said the mycologist Alan Rockefeller in a conversation, "because no mushrooms have aluminum-colored spores.") Cover the mushroom with a glass or bowl and leave it to rest overnight. The next morning, the gills will have dropped their spores, often in beautiful patterns. If you have picked what you thought was a *Psilocybe* but it produced a rust-colored spore print, then you might have just saved yourself from an entanglement with the deadly galerina.

Wild mushroom hunters—and eaters—face the possibility of a mistaken ID every time they go into the woods (well, not so much the distinctive *Amanita muscaria* but certainly among the LBMs). But at the same time, said Maya Shetreat, a neurologist and author of *The Master Plant Experience*, "hunting mushrooms is a kind of medicine," from forest bathing, "which decreases cortisol production and improves focus and executive functioning and subjective feelings of happiness," to lateral eye movement, which "engages both hemispheres of the brain, getting you into a parasympathetic state"; even forward walking, she said in an interview, is beneficial for the feeling of well-being. It works for me, and I think for my silent companions in the woods, too. I can head out with a group of mushroom hunters, people I don't know at all, like the crew from Long Beach, and find a kind of quiet camaraderie. We don't follow one another; we don't pick someone else's patches. We basically stay clear of one another, but our mental states are aligned. For us, mushroom hunting is a form of self-soothing, a kind of therapy where we all chip in gas money to get there.

6

Cultivating

Despite its pleasures, and maybe because of the potential poisoning issue, many people who use psychedelic mushrooms, especially those who use them frequently, like microdosers, don't become mushroom hunters. Instead, they end up procuring their mushrooms from someone who has grown them or become growers themselves. Mike Maki, who turned eighteen the day Martin Luther King Jr. was assassinated, used to dry and sell the mushrooms he picked, and then, eventually, he figured out how to grow *P. cubensis* at home. Ironically, at the same time the customer base for Maki's mushrooms was increasing, so was the number of home cultivators. Today, the cost of a pound of *P. cubensis* is half what it was in 1980. As Maki, who is into growing functional and culinary mushrooms these days, says, "There are some pretty darn smart people clustered around this theme of mushroom growing."

The world of magic mushroom cultivation is rich in brilliant if seedy characters, sincere off-the-gridders, rogue microbiologists, and seekers of heavenly access. Their numbers have increased in parallel with the spread of communication technology, from pamphlets and books to the internet. And word of mouth. Always word of mouth, because while magic mushrooms are illicit to some, scary to others, holy to many, they are simply illegal to the Feds (and others). The banning of psilocybin in 1971 pushed the early growers underground but didn't wipe them out. They have continued to share tips and proliferate in number as grow techniques become simpler and simpler.

Today, magic mushroom growers are not unlike the orchid crowd, fascinated by new cultivars and obsessed with creating new morphologies. Mushroom cultivators are caregivers, if sometimes zany ones. They love to watch their mushrooms grow, and that's evident because there are plenty of cultivators who grow far more mushrooms than they could ever consume, and most home growers, in my experience, are not into selling. Many abide by a kind of code that dictates they only share the mushrooms. They grow purely for the sake of growing; enthusiasts call it a "fantastic hobby," albeit one that could get you a felony conviction.

As of this writing, psychedelic mushrooms are subject to a variety of regulations in the USA, but for the home cultivator, the situation needs some explanation. In most states, it's legal to buy spores for study, but not for germination. That's because spores don't contain any psilocybin or other prohibited psychoactive alkaloids. Nor does the fungus that grows from the spore seem to produce much psilocybin—though there is some. I've seen the evidence when I've kicked open piles of wood chips along parkways in Seattle to reveal pale blue fungal threads, indicating the presence of psilocybin. But the alkaloids seem to be concentrated largely in the mushroom, which is what critters like the mycologist Bryn Dentinger's slugs are most likely to consume.

Psilocybin and psilocin are banned by the federal government. Even researchers need to get special permission from the DEA to use the drug in clinical trials. The possession of mushrooms is also banned because they are considered a container for the drug. A patchwork of laws in various states and cities have decriminalized possession of *Psilocybe* mushrooms for personal use, but that doesn't eliminate the possibility of arrest by the Feds or in any jurisdiction where the mushroom is illegal. So any potential cultivator simply must assess the legal risk.

All sorts of people seem to be willing to take that risk, from socialites to stay-at-home dads. A young fellow who was ringing up my bill on a piece of jewelry repair showed me photos on his phone of a miniature forest of mushrooms he grew in his Brooklyn kitchen, a plastic bin erupting with *Psilocybe cubensis*, white-stalked with brown caps. For a month or so, I tried to make an appointment to come see him and his mushrooms, but as he'd just harvested one batch and was starting another, I had to wait—it takes four to six weeks to get from inoculation of the growth medium with live spores to the flushing—or fruiting—of mushrooms. In the meantime, he introduced me to another man, living in upstate New York, who started growing mushrooms for personal use during the COVID pandemic. "I looked online and found the most elementary school of growing, ordered all the materials, and it worked out perfectly." He was talking about PF Tek, a simple technique for what the mushroom growing community calls *noobs*, or first-time growers. I asked if I could come to see his mushrooms, and he said, "Why don't you just grow your own?"

He sent me a link to PF Tek for Simple Minds, a web page that explains how to cultivate *Psilocybe cubensis*, decorated with illustrations of gnomes resting against *Amanita muscaria*. The *PF* in PF Tek stands for *Psylocybe Fanaticus*, the online name of the late Robert McPherson, a jazz musician and creator of the grow technique. (He was also a terrible speller—*Psilocybe* is spelled wrong, but once the company name was established, he had to stick with it.) *Tek* stands for *technique*. PF Tek's beauty was and is its simplicity. After reading through the instructions a few times, I felt like this was doable. My first move: purchasing a spore syringe. The spore syringe was a significant improvement when it came to home cultivation because it allowed growers to skip the stage of germinating spores on agar. I think at this point, an outline of how mushrooms grow might be helpful.

A Little Basic Mycology

Psilocybe mushrooms are basidiomycetes, which means their spores—millions of spores—attach to tiny fingerlike appendages that hang from the gills under the caps. (Actually, the gills are called *lamellae*, the mushroom cap is called the *pileus*, and the stalk is called the *stipe*: fungi are in a kingdom of life all their own, so they have special names for their parts that are distinct from plants. That said, nonmycologists tend to use plant terms when describing mushrooms.) When the wind blows or animals brush by, the spores disperse into the air like pollen. If a spore lands in a habitat with the kind of food it likes to eat, it will germinate and grow a cell at a time in long, tubular threads called *hyphae*. These are microscopically thin, about ten times thinner than an average human hair. The hyphae of *Psilocybe* fungi grow by branching and rebranching in every direction they have food to grow into, each hyphal tip (and even along its length) a food absorption point. This mass of threads is called *mycelium*. When you kick open a rotten log, that white cotton candy–like stuff is mycelium. It's not always visible—there must be lots of it—but soil is riddled with fungi. One teaspoon of healthy soil contains so many hyphal threads that if you laid the hyphae end to end, the thread they form could be six miles long.

A *Psilocybe* mycelium contains a nucleus in each hyphal cell. As the hyphae grow, the nucleus divides, and a cell wall forms between them, creating individual cells with a nucleus in each. Having a nucleus in each cell allows the mycelium to continue to grow even if it is broken apart: the know-how to grow is in each nucleus. Mycelial regeneration is like the broom in *The Sorcerer's Apprentice*: you can take an axe to the broom, but the individual pieces just grow into more brooms. This ability to regenerate is why, according to the mycologist Mark Miller, a tiny fragment of *Agaricus bisporus* mycelium the size of a watch battery and the weight of a housefly can grow into a mycelium large enough to produce one hundred thousand pounds of white button mushrooms. Growers call this stuff *spawn*, from the Latin verb *expandere,* to spread out.

A mycelium can keep growing like this as long as it has food to consume, but it won't produce mushrooms. That's because the nucleus in its cells has only a single copy of each chromosome, and you need two copies to make a mushroom. That's accomplished when one mycelium encounters and merges with another mycelium of the same species whose nuclei contain a compatible chromosome, something like two streams coming together to form a river. The result is a mycelium with two nuclei in each cell that contains the genetic material from both original mycelia.

It is from this secondary mycelium that a mushroom forms.

First, hyphal threads that are part of the mycelium wind together and ball into a knot, like a pill that forms on a sweater, differentiating into parts that will eventually become the stalk and cap. Another word for these knots is *primordia*. It's the mushroom version of a very young flower bud.

When it rains, the mycelium gathers moisture from the ground along with essential nutrients like phosphorus and nitrogen and pumps it into the primordia. That's why mushrooms seem to appear suddenly after a rain: the mycelium inflates the primordium with water, and it swells into a mushroom. Spores from *Psilocybe* species (and many others as well) are formed in the basidia of the mushroom—those tonguelike appendages that hang from the gills—and it is there that the true sex happens: the two nuclei in the cell fuse into one that contains the full suite of genes. That cell then divides, mixing up the genes and reducing the number of chromosomes in each future spore to one set, and the cycle starts again.

The Ways We Cultivate

It took a half century of people tinkering with methods and substrates and the challenges of cultivating in one's kitchen before bins of magic mushrooms could be grown under the narrow beds in college dorm rooms all over the country. The French mycologist Roger Heim and his collaborator Roger Cailleux accomplished the first successful cultivation of *Psilocybe* in the late 1950s. Heim used the same technique that was applied in the early days of white button mushroom (*Agaricus bisporus*) cultivation: Nineteenth-century mushroom growers collected

A. bisporus mycelium in the wild and transplanted it to beds of composted manure, where it was allowed to proliferate. The mycelium was collected, dried, and shipped to growers around the world, who would then scatter the mycelium into composted manure, add a layer of moist material called *casing*, which aids in moisture retention, and wait for the mushrooms to erupt. Heim used mycelium collected in Mexico and successfully flushed *Psilocybe mexicana* on straw compost with a casing layer of sand and chalky earth. The mycologists grew a few other species, including *P. cubensis*, which flushed on cased horse manure compost. Few knew about their success, however, as the results were not published in English until 1977, in the now-rare *Magic Mushroom Cultivation* by the late Steven Pollock.

Back then, *Psilocybe* cultivation was primarily a scientific endeavor, requiring microbiological skills to manage the problem of sterility. And synthetic psilocybin was rare—much rarer than LSD, as it is both expensive and difficult to make. But cultivation and synthesizing techniques eventually started to seep into the counterculture. The pamphlet *The Psychedelic Guide to Preparation of the Eucharist*, edited by Robert E. Brown and his associates in the Neo-American Church* and published in 1968, laid out the steps for growing *Psilocybe* mycelium in culture and making synthetic "psilocyn" from it. The pamphlet is quite technical, the antithesis of what you'd expect from the kaleidoscopic cover art and the title. (It's also costly: around $500 from used booksellers.)

By the early 1970s, a cultivation technique that anyone

* The Neo-American Church was a religious organization based on the use of psychedelic drugs, founded in the mid-'60s by the psychologist Arthur Kleps (1928–1999).

could manage started to proliferate in Seattle, when University of Washington students realized there were magic mushrooms growing in landscaping wood chips around campus.

Noccers

Most of the time, the mushrooms find new habitat when the wind blows their spores around. In the case of urban psychedelics like ovoids and wavy caps, the mushroom mycelium may be dispersed via the wood chips themselves. If the source wood chips are infected with the fungus, then every shovelful is a potential colony. It's the same model as commercial growing, where a jar of grain is inoculated with fungal spores; the spores grow into mycelium, and the mycelium is divided among multiple jars of grain, where the mycelia continue to grow. Likewise, a *P. cyanescens* mycelium that infects a seminal wood chip supply can hitchhike on wood chips to new habitats, where it continues to increase exponentially. As it turns out, the University of Washington's landscaping contractors unknowingly were cultivators of psychedelic mushrooms around campus. The unintentional proliferation of magic mushrooms on the UW campus and in Pacific Northwest cities more generally has been augmented by folks who deliberately transplanted mycelium-colonized wood chips into new beds. And they still do. For some, this is a hobby; for others, a calling.

Noccers, short for *inoculators* (and pronounced *knockers*), are rogue cultivators, mycohooligans who are proficient amateur (and sometimes professional) mycologists with an antiestablishment streak and a fine sense of irony. I know a small group of noccers who operate in the Seattle area. They scope out newly landscaped locations, following tips from their mole inside a landscaping company with a parks department contract, and promptly inoculate them with *Psilocybe* mycelium. I asked one noccer, I'll call him Pooch, if I could see where he'd inoculated over the last year, and he texted me a GPS coordinate. And then another, and another—all told, around thirty coordinates that I spent a day tracking down. My companion, the writer Langdon Cook, and I cracked up at the sites: mostly drive-up-able edges of parks, housing developments, Pooch's parents' condo, and churches. One site was inside the batting cage of a Little League diamond. That was unintentional, Pooch insisted. He'd inoculated a nearby pile of wood chips with ovoids—a gregarious, nitrogen-greedy species that grows vigorously but depletes the wood chips of their nutrients in a year—and someone *else* moved them into the batting cage, where they erupted in tremendous numbers at the beginning of the spring baseball season.

Noccers have at times filled balloons with a slurry of spores and water and chucked them onto wood chip dumps awaiting distribution by the city, and then, when they find mushrooms growing, they capture some of the myceliated wood and move it elsewhere: to the landscaping around police stations, the courthouse, along power lines, behind ornamental bushes in city arboretums and botanical gardens "so you aren't easily caught picking." One day in Seattle, I racked up nearly twenty thousand steps—almost ten miles—just wandering around nocced hardwood chip sites with Pooch. It was spring, and we were looking to see if the ovoids were up. I collected a bagful of the fruiting mycelium and brought it to a dinner party on bosky Bainbridge Island that night (where it was proclaimed the Best Hostess Gift Ever). A few people took a taste. Fresh, wet ovoids are moderately potent, and about thirty minutes into the evening, it wasn't totally clear who was cooking dinner, but ultimately, everything came together, merrily.

Noccers employ a few different techniques. The simplest is collecting the "stem butts" of picked *Psilocybe* mushrooms—that little clump of soil and mycelium at the base of the stem—and packing them into a plastic bag to let the mycelium colonize all the soil debris. The butts can then be tossed onto a pile of (ideally) year-old hardwood chips. "Not cedar and not dust," warned one noccer who did a noccing presentation at a PNW foray I attended. "Dust chokes them out." Another technique is to make spore prints of psychedelic mushrooms on a piece of aluminum foil, spritz the spores with water, and add that slurry to a bottle of water with a spray top. A spore print made on a paper towel will do, too. Just dampen the towel and throw it into a pile of chips. The mycelium gets going eating the paper first. "Or," said our presenter as the noccer demo came to a close, "just rub your shoes with nasty mushrooms," those that are too far gone to consume or dry, and go for a walk. You will essentially be propagating these mushrooms on a landscape scale.

Pooch admitted to hundreds of inoculations. "I did around fifty last year," he said, "and I've been doing it for about seven years. I consider it a public service." There are probably psychedelic mycelia in Seattle that are fifty years old, their genotypes moved annually to new habitats over the generations. That's given rise to phenotypes. Ovoids, for example, flush in two varieties: what Pooch calls *chunkies* and *bells*. They are differentiated by their morphology, but they also have slightly different psychedelic effects. "The chunkies are more mellow," said one slender, feline noccer we hunted with, a young man whose day job is working in cat rescue. Moving these phenotypes around is a way to keep them going. Indeed, that was the main interest of Brandon Fogg, a veteran noccer who works in the aerospace industry. When he found a species flushing in a new habitat (say, potting soil versus grass), he would collect the spores, clone them, and reintroduce the phenotype into new locations. "Like a breeder would do," he told me, "but my lab was the outdoors."

Noccers tend to be foragers of many foods; psychedelics—or, in their preferred terminology, *actives*—are just the gateway to their greater involvement with wild foods, ecology, and nature. And it *is* practically a lifestyle: inoculating must be done annually if you want to keep the mycelium going, because as soon as the wood chips are totally degraded and have returned to a soil state—usually two to three years—the flushing is over, and a new habitat must be found. As one noccer said about the wood chips of urban Seattle, "The mushrooms are either there, were there, or will be there."

That's the case of one such spot in New York City. I reached out to a mushroom hunter, John, who told me a *P. cyanescens* spot near Inwood Hill Park was always on the verge of petering out because no new wood chips were laid down. John went on to share detailed descriptions of other nooks and crannies in New York City where *P. cyanescens* grows but admonished me not to mention any of this "to [fellow mushroom hunter] Dennis because I'm sure he will have a conniption."

The divide between the students who first transplanted naturally inoculated mycelium and the science lab was bridged in 1976 with the publication of *Psilocybin: Magic Mushroom*

Grower's Guide: A Handbook for Psilocybin Enthusiasts by O. T. Oss and O. N. Oeric, pseudonyms for Dennis and Terence McKenna. Their book adapted breakthroughs in the commercial mushroom industry to the home kitchen. It sold over one hundred thousand copies in its first five years of publication. *Magic Mushroom Cultivation* by Steven Pollock came out the next year. Pollock experimented with a variety of substrates, ultimately determining that brown rice was the cheapest and easiest to use. Both books recommended using mason jars filled with a substrate and sterilized, then inoculated with spore or, even better, pure culture—spore germinated on agar—but they utilized different substrates, which could affect not only flushing rates but potency as well.

More tweaks were made. For example, in 1983, the mycologist Edmond R. Badham found that damp brown rice with a casing layer of moist vermiculite did the trick. That same year, Paul Stamets and J. S. Chilton published *The Mushroom Cultivator*, which explained how to grow fifteen different species of mushrooms, including *Psilocybe*. Experimentation carried on over the next thirty years, with different substrates, including birdseed, compost, cow or horse manure, worm castings, and, my favorite, Ben's Original Ready Rice. Along the way, the technique for growing magic mushrooms became simpler and simpler, involving fewer steps, utilizing ever more accessible ingredients (no more "slightly aged manure, not young, not old"–type instructions), and in the process creating an ever-larger cohort of home growers. One problem remained, however: the inoculation of the substrate with spore. That's when the substrate is especially vulnerable to contamination by other microbes.

And so, so much can go wrong.

Enter Spore Syringes

Contamination is probably the biggest hurdle to growing mushrooms. The air is loaded with bacteria and fungal spores that can infect the substrate during inoculation, during mycelium colonization, or during the flushing stage. Contamination can be triggered by overnutrition or undernutrition of the substrate; too much water or too little; too much heat or too little; conditions that favor the presence of one contaminant or another; or just bad luck—all of it dependent on whether these contaminants can get into your substrate in the first place. Mushroom growers would be well advised to take the traditional Chinese medicine approach to health and avoid contamination because it is a lot more difficult if not impossible to take the Western approach and kill the contaminants, which would end up, in the case of your mushroom crop, killing the patient. For that reason, magic mushroom growing didn't get easy for someone like me until the innovations of Psylocybe Fanaticus were first published in *High Times* magazine in 1991. It described how to use a syringe to inject spores in liquid solution directly into a substrate composed of sterilized brown rice flour, vermiculite, and water. Home cultivation became even more accessible with the development of syringes prepacked with *P. cubensis* spores in solution.

P. cubensis is the easiest psychedelic mushroom to cultivate. But of course, it started out wild and is still collected in the wild. It was first collected by Franklin Sumner Earle (1856–1929), an American mycologist who specialized in the diseases of sugarcane, which brought him to Cuba, where he found the mushroom, hence the name *cubensis*. It is larger than many species, with a golden-brown cap about the size of a vanilla wafer and a white stem that bruises blue.

P. cubensis may be native to the Americas on herbivore dung, or it may have immigrated from sub-Saharan Africa, where its close relative *P. natalensis* thrives. Either way, the mushroom likely dispersed by following the geographic patterns of cattle farming in semitropical areas prone to warm summer rains. It is, as Dr. Dentinger says, "self-domesticating," proliferating in the aromatic wake of domesticated herds that are themselves the product of human domestication.

Today, *P. cubensis* can be found from the American southeast to Mexico, Cuba, southeast Asia, India, Australia, and northern South America. Like *P. semilanceata* and pastures, and *P. cyanescens* and wood chips, *P. cubensis* has thrived in the wake of our preferences, in this case, our love of hamburgers. In Mexico, it is known as San Isidro de Labrador, named for the patron saint of the fields (or agriculture or the plow), though in Mazatec, the name translates to "Divine Mushroom of Manure." Mushroom hunters have collected *P. cubensis* in all these places. They have made spore prints and harvested the spores for cultivation, and some of those spores are available for sale on the internet today.

Lots of companies sell magic mushroom spore syringes, as they are legal, at this writing, in most US states (though not California, Idaho, and Georgia), Canada, most of Europe, and in Central and South America. I picked a company that's been in business for many years and whose website was easy to navigate. It sells glass cover slides and microscopes and syringes for a variety of *P. cubensis* cultivars, including Tidal Wave, rated at press time highest in psychoactive molecules. It also sells a range of psychoactive species, including *Gymnopilus luteofolius* and

Panaeolus cyanescens, culinary mushrooms like maitake (*Grifola frondosa*), and functional mushrooms like reishi (*Ganoderma lucidum*). There is a lot of choice.

That's why it is important to look for the Latin binomial when buying spores. Different species have different growing requirements. It's a little like buying flower seeds—they don't all like partial shade, and if that's all you've got in your yard, you would be out of luck if you didn't understand the particulars a species requires to grow. That said, most psychoactive spores for sale are *P. cubensis* because they're so easy to grow. But there are many cultivars.

Cultivars

Cultivars, or cultivated varieties, are grown from spores that contain a defined mix of genes. Selective breeding of the mushrooms and then propagation of those spores can create distinct characteristics. While there can be interbreeding between cultivars of a single species like *P. cubensis*, it's rare between two different species. They just don't have enough genes in common for that to work. Strains, on the other hand, are cultivars that are maintained by cloning. For mushrooms, this means they have been grown out from a piece of mushroom tissue. This works because the tissue of the mushroom is still-living mycelium. (If you harvest spores from a strain, the mushrooms could express any combination of the genetics present and no longer be identical to the generation before.)

Most cultivars in circulation are *P. cubensis*. Cultivars have emerged that alter the size of the mushroom, the cap color, gill production, flush vitality (how many times a single substrate will produce mushrooms), colonization speed, and psychoactivity. There are cultivars that are named for their appearance, like Albino A+, which is a large all-white mushroom, or Penis Envy, a mushroom whose cap is so tight you can't really see the gills (and with impressive psychedelic provenance: the strain was isolated by Steven Pollock* from spores he received from Terence and Dennis McKenna). There are whimsical cultivars like Aztec God, and cultivars that make promises, like Jedi Mind Fuck. There are cultivars named for various cultivators, like P. Menace and Lizard King, and then a whole bunch of cultivars named for where the original specimen was collected: Burma, Cambodian, Chilean, and Chitwan from Nepal—over one hundred cultivars in all. While what happens in your brain is mostly a matter of dosage, the percentage of psilocybin and psilocin in different cultivars is key to branding. Increased psychoactivity makes a cultivar more desirable, both for sale to cultivators and those in the spore-trading community.

One of the places where spore traders can find out about the potency of new cultivars is the Cups. These are like the Westminster Dog Show, only for mushrooms. Growers submit their cultivars, which are measured for their psilocybin and psilocin content, along with other relevant data. The Psychedelic Club of Denver hosts the Psychedelic Cup, and Hyphae Labs in Oakland, California, hosts the Hyphae Cup (formerly the Psilocybin Cup), and in both cases, they publish the submission data on their websites. Over one hundred participants contributed mushrooms to Hyphae Labs' Spring 2022 Psilocybin Cup, with cultivators going by names like DicDanger and MycoCowboy,

* Pollock, a medical doctor and the author of *Magic Mushroom Cultivation*, owned a spawn-making company, Hidden Creek. He was shot in his home in Texas in 1981. I reported on this in *Mycophilia: Revelations from the Weird World of Mushrooms* (New York: Rodale, 2011).

and submitting entries like Stingray and DeathStar. The Cup is organized into categories based on use: spiritual, therapeutic, recreational, and microdosing. The spiritual champion that year was APE BC, with 25 milligrams of psychoactive tryptamines per dried gram. The Humble Tortoise's PE7 contained about one-third the amount: 7.3 milligrams per dry gram. Keeping in mind that both these cultivars are the same species, it's clear there can be a pretty big difference in strength from one cultivar to another. According to Alan Rockefeller, 2 percent tryptamines, or the equivalent of 20 milligrams per dried gram, is very, very potent; 1 percent is very potent; 0.75 percent is still strong; 0.5 percent is moderate; and 0.25 percent is weak. But you can't count on consistency when it comes to these cultivars (unless they are clones). Between variety among specimens and variety of growth conditions, yesterday's APE BC might be today's PE7.

The PF Tek

Most growers who are really committed to the hobby will start their own cultures from spores, on agar in petri dishes. It's the next level of cultivation, allowing the cultivator to develop cultivars of their own, with all the attendant naming and bragging rights. It also requires more practice and a higher degree of sterility consciousness as well as equipment like pipettes and inoculation loops. The beauty of the spore syringe is that it significantly reduces the possibility of introducing spoilers into your sterile jars. Of course, the inside of the syringe is not sterile, so there's no guarantee there aren't mold spores floating around in there, too, but who are you going to complain to? You're not supposed to be germinating in the first place.

I ended up buying one 10 cc syringe of *P. cubensis*, the strain Golden Teacher, the equivalent of choosing vanilla in an ice cream parlor. I placed the order over the phone with a cheery salesperson. It was like buying pillows at the Company Store, except they take cryptocurrency. A few days later, I received a slim box from a return address I didn't recognize. Inside was a large plastic syringe, glass slides, and a thick needle in a plastic sleeve. There are, thankfully, videos on the internet showing how to attach a needle to a syringe. The cavity of the syringe was filled with a thick yellowish liquid, water made heavy with agar (an innovation Psylocybe Fanaticus claimed credit for), so the spores do not stick to the sides of the barrel. I could see suspended clumps of purply black gunk that needed a few good shakes to break up and distribute.

I am a home canner, so luckily I have some experience with the challenges of producing sterile jars, and I have the basic equipment: a pressure canner (mine is a ten-quart All American with clamping locks, a weighted gauge, and a dial gauge) and a cabinet full of empty mason jars and rings. Home canners focus on killing pathogens that are present in our foods and keeping new ones out of our jars. We are kitchen microbiologists. And such knowledge is handy when it comes to sterilizing mushroom substrate.

According to the authors of the *Psilocybin Mushroom Handbook*, *P. cubensis* will grow on any substrate that is sufficiently rich in carbon and nitrogen, including cereal grains, grass, corn, wood, even cardboard. It's a highly competitive environment, with lots of microbes interested in that same food, which is why sterilizing the substrate before introducing the spores is so important. To make their substrate, the PF TEK calls for moistening a combination of vermiculite—which I self-consciously purchased at a local gardening store, since I have no idea what vermiculite is good for other than growing magic mushrooms—and brown rice flour, loosely packed into half-pint mason jars. (I later learned that soaking the substrate in water at room temperature overnight before packing it in the jars can encourage any bacterial endospores to germinate and as a result be more vulnerable to destruction when you sterilize the jars. Oh well.) The jars are covered in two layers of aluminum foil, then pressure canned to sterilize the vermiculite-flour combo.

The instructions call for one inch of water in the bottom of the canner—less than we normally add for canning foods—and fifteen pounds pressure per square inch (psi). That's higher than we use for pressure canning raw meat at sea level. The time stays the same whatever your altitude—forty-five minutes, about what it takes to process a jar of sliced button mushrooms. What that means—and what happened to me—is that the water in the canner all but evaporates. Without ample water to turn into steam, the pressure won't build, and your substrate doesn't get sterilized. It makes more sense to me to put a rack in the bottom of the pot to lift the jars and then add about three inches of water, like you do when pressure canning green beans. On the other hand, the message boards are full of people getting drunk and forgetting their jars, and thinking they'd burned the substrate because all the water in their canner evaporated, but then posting photos of jars exploding with mushrooms. It seems there is no definitive technique for growing mushrooms. They just do what they want to do.

Once the jars cool, it is time for the tricky part: inoculation. Since COVID, many of us have become aware of how clean a workspace must be to be free of microbial contaminants. So I gloved up, put a plastic shower cap over my hair, and repeatedly wiped down my surfaces with alcohol. The authors of the *Psilocybin Mushroom Handbook* also suggest I shower and invest in some mental hygiene: be calm and play soothing music. They specifically advise against music by Karlheinz Stockhausen (1928–2007), a German electronic composer whose music I promptly downloaded onto my phone. They also recommended clearing the house of pets and advised paying cash for everything, an impossibility these days. Nor did I worry about these overblown instructions. I mean, they also suggested I repaint my kitchen.

After removing the top piece of foil, I sterilized the needle by holding it over the flame of a lighter until it turned red—exactly like we used to do with safety pins when we pierced our ears in high school—then plunged the needle through the foil that remained, administering a 1 cc squirt at twelve, three, six, and nine o'clock before replacing the second foil lid. In all, I inoculated six jars this way. More and more carbon built up on the needle with each re-sterilization, but no need to worry about that; it does not contaminate anything.

The next step is incubation. This is basically a water bath, which I created by placing a fish tank heater in the bottom of a plastic bin, adding water, then putting another plastic bin in with my sterilized jars that I'd wrapped in a towel. I had to add weights to stabilize the upper bin—the whole thing a kind of rickety rig—and then went away for the weekend. When I came back, nothing had happened. I checked the sides of my jars for growing mycelium every day for ten days until finally I started to see tiny white patches. Spores are microscopic, and germination is, too. These cells are growing one at a time, so it takes a long time before the fungus is evident to the naked eye. Just as eagerly as I was looking for evidence of a healthy mycelium, I was also looking for evidence of contamination. A common contaminant is *Bacillus* bacteria, which can thrive in your grain jar if its hardy endospores survive the sterilization stage. *Bacillus* looks like patches of brown mucus—hence the nickname "wet spot"—and smells, according to the experts on Shroomery, "like rotting apples, dirty socks or burnt bacon."

All kinds of gross things can grow in those jars: cobweb mold (too much humidity), green mold, cinnamon-brown mold, red bread mold, yellow mold, black whisker mold, blue-green molds (the *Penicillium* spp.), pin, plaster, and flour molds, to name a few. Growers might contend with La France disease (a virus), mummy disease, wet bubble, dry bubble, fungus gnats ("The person who figures out how to get rid of gnats will make a fortune," said one former cultivator), mites, and abiotic disorders like browning, malformation of cap or gill, scaling, stroma, and weepers. The magic mushroom cultivation chat groups are littered with subject lines like "Seeing some white goo on these mushroom heads." A group like r/psilocybingrowers has twelve thousand members, constituting a substantial resource should spots of green snot appear in the jars.* In most cases, once infected, the "cake" (the colonized substrate with whatever contamination is present) must be discarded. However, that doesn't mean all is lost. Numerous cultivators have shared photos of contaminated substrate thrown into gardens and even potted plants where, after a while, the mushrooms have made a comeback.

One of the most common questions noob growers ask is "Why is it taking so long for my mycelium to colonize the jar?" One reason may be gas exchange. Fungi are aerobic—they ab-

* An endearing aspect of these chat groups is the dichotomy of posters' kooky handles, say, ShroomPirate, and their often very technical advice, like identifying a microbial contaminant by its scientific name.

sorb O_2 and expel CO_2. My guess is my jars colonized slowly because there wasn't ample gas exchange. That can be remedied by allowing tiny amounts of oxygen in—and keeping spoilers out—by cutting slits in the foil and covering them with a micropore tape. The other reason may be heat. I had no idea if my Petco thermometer was accurate, and I was a little suspicious because it didn't vary by one degree the entire time.

I was surprised by how much I cared for those jars; I examined them daily—twice a day, even—to see how they were doing, as if they were pets. When I first noticed those little patches of mycelium, I cooed. After a month, four of my six jars were mostly colonized with mycelium. Two never took off, but I kept them anyway, just in case they weren't dead. In the jars that were colonized, I could see whiskery threads reaching up into the headspace of the jar, but I had to wait until the entire jar was colonized, thick with mycelium to the point where no brown substrate could be seen. This is important because once the fungus has consumed the substrate, it has an evolutionary impetus to flush and get its spores out to find a new habitat, just as it does in the wild. And while I waited, I checked in regularly with the cultivation chat groups, to be sure my jars were on track, but also because I became fascinated by the zany characters who populate the online cultivation world.

Cultivation on the Internet

Home cultivation really took off when the PF Tek was published online in 1996 on alt.drugs, a recreational drug discussion forum founded in 1992 ("a century ago in NetYears," point out the editors of Erowid.org, maybe the premier website for candid info about drugs). Since then, it has been "downloaded, printed, distributed and practiced millions of times." As Psylocybe Fanaticus wrote with classic cultivator hubris, "I am responsible for the greatest proliferation of magic mushrooms in the history of the human race." The alt. hierarchy was founded by civil libertarian and early (like, fifth) employee of Sun Microsystems John Gilmore and Brian Reid, a computer scientist and developer of protection techniques that went on to become the first internet firewall. They built it as an alternative network for people who wanted to talk about subjects deemed taboo by Usenet, an early network that contributed to the internet.* A year later, a teenager named Ythan Bernstein built Shroomery over his summer vacation, and in 2001, Mycotopia was founded by Hippie3 (*3* because he lost Hippie1 and Hippie2 to other users), a former getaway driver.

These forums were characterized by constant drama and disputes over credit for growing innovations. Today, they are rowdy spaces with layered conversations about species, cultivation, and contamination; loaded with irreverence, libertarian ethos, flames, boasts, and camaraderie, and lots of funky homemade GIFs. Should you have the stamina to sort through the posts, you will encounter a wealth of information and misinformation alike (no, says a doctor friend of mine, there's no evidence you can get a bacterial infection from sniffing a contaminated jar of *Psilocybe* mycelium). Clare, a fellow amateur mycologist, has been growing magic mushrooms "since the dawn of

* Besides being the original landing page for PF Tek, the site covered a range of subjects, notably, the dried-banana-peel hoax and smoking-peanuts-to-get-high hoax, both of which were mentioned in *The Anarchist Cookbook*. As the moderator of the site wrote, "If you have the Anarchist Cookbook and are actually dumb enough to follow it without doing some more research, you are likely to become a classic example of evolution in action."

the forums" (that is, about twenty years). "Shroomery would lock threads if they were not sure the content was accurate," he said. But Mycotopia was more freewheeling, allowing for more experimentation ("Let's try colonizing [*Psilocybe cubensis*] on hominy from a can!"), and many of the scene's luminaries, like the ethnomycologist John Allen, for whom *Psilocybe allenii* was named, checked in. More recently, mycologists like Alan Rockefeller have provided insights on the genus to the public via Shroomery, Mushroom Observer, and iNaturalist.

The forums spawned a thriving subculture of home cultivators and DIY geneticists working to advance cultivation techniques. Innovations like the Dunk Tek, which calls for dunking the mycelium cake in water to encourage rapid flushing, and discussions about, say, the benefits of adding nitrogen at various stages in cultivation to boost the mushroom's production of alkaloids flourished in that environment. They also led to pure silliness. Roger Rabbit (a.k.a. Mark R. Keith) was well known for growing magic mushrooms in the oddest places, from a Bible into which he slipped plates of inoculated agar between the proverbs to a bra he packed with inoculated substrate and, when it fruited, modeled.

It was in this ecosystem that *Psilocybe cubensis* cultivars expanded. Psylocybe Fanaticus (or his grow partner) discovered the albino cultivar because they were growing under UV lights. Cultivators crossed mushrooms that exhibited random mutations like albinism with other cultivars to create novelty varieties like the Albino Golden Teacher. This is citizen science at work; DIY mycolabs pushing the science forward and sharing it via the internet, mixed with a healthy dose of capitalism. Breeding a mushroom with high alkaloid production or that was just weird-looking could bring infamy in the forums, not to mention sales of spores and related merchandise. Novelty spores like Tidal Wave can cost eighty to one hundred dollars for a syringe. The fellow who bred TAT (True Albino Teacher) sells pins, beanies, and T-shirts with the cultivar's logo. And of course, there is plenty of puffery. Take the cultivar Sacred Sun, a "masterpiece . . . endorsed by spiritual healers worldwide . . . [a] witty, yet powerful strain," which sounds to me like the psychedelic version of a sommelier trying to sell an overpriced bottle of wine.

My cultivar, the humble Golden Teacher, had mostly colonized the jars after seven weeks, so I decided it was time to give birth. I emptied the plastic buckets and poured a half inch of pearlite into the bottom of one, spritzed it with water, and then, kneeling over the tub, set about birthing. I removed the foil cap, gave the jar a couple of good shakes and then a hard smack on the bottom, and out the mycelium cake slid onto the pearlite. A few refused to release, so I wiped a knife with alcohol and broke the suction. Of my six jars, four birthed; the other two showed no signs of fungus. I covered the top of the bin with a clear plastic bag and left it in indirect light. I had plans to go on vacation in a week, so I was hoping for a fast flush.

In the end, only two out of my original six jars produced mushrooms. Once the first brown-tipped pins appeared, the mushrooms took about seven days to mature. I knew they were ready to harvest when their caps opened, before they dropped purply brown spores all over the cakes. There were some little aborts—tiny mushrooms that grew at the base of the larger ones—and I picked those, too, by cutting them carefully with a sharp knife. A tomato farmer friend of mine considers aborts a specialty. He dries bags of the nubbly blue mushroom stubs, no bigger than a pencil eraser, which, despite their small size, can provide a mighty punch.

Mushrooms that are not eaten raw need to be preserved, and quickly, too, as they are prone to rot. The most common way to preserve is drying in a food dehydrator. The dirt is cleaned off with a brush and the mushrooms dehydrated until cracker crisp, usually up to twelve hours at 125°F. As with culinary mushrooms, if they are washed before dehydrating, they will take longer to dry out, and the more time they take to dry out, the greater the chance mold will get a foothold. The herbalist Christopher Hobbs pointed out that the psilocybin in the mushrooms is sensitive to heat and light, so after drying they should be kept in a cool, dark place, or even in the freezer. Freeze-drying may be the best way to keep them over time.

I ended up with six mushrooms about five inches tall, about three dried grams, not including the aborts. Nothing to brag about, but enough to go places you haven't been. For another week, I waited for the cakes to flush a second time, but nothing happened. That doesn't mean the process was over, however. My batch might have started from cultivation, but I hoped I would end up hunting them, because a few days later, we took a little walk to a local park and, when nobody was looking, chucked the myceliated substrate on a bridle path. (*P. cubensis* likes horse manure.) I'll be checking that spot for mushrooms next year.

7

Clinical Trials, Retreats, and Private Guides

Both hunting and growing magic mushrooms require a learning curve, and many people aren't going to take the plunge into amateur mycology. What more folks do is acquire the psilocybin experience through a clinical or commercial entity. And the internet is loaded with opportunities to trip, both legal and not.

That's because many folks are psychedelic curious: about mind expansion, spiritual growth, self-improvement, and the potential of therapeutic benefits. The good news is an industry has grown up to provide psychedelic services, from retreats, where participants may experience the mushroom in a safe environment, to individual guides who offer their services at a client's home or somewhere else, even virtually. And as the number and size of clinical trials increase, one can expect more opportunities for people to participate. The bad news is clinical trials are still small and hard to get into, and there is no regulation of the psychedelic industry. Which means there is no guarantee of successful outcomes and, in many cases, no redress should any violations occur.

The State of Psychedelic Research

For the last decade or so, there has been an accelerating drumbeat of publicity about psychedelic drugs. As more and more people hear about the potential of psilocybin to counter depression and anxiety, researchers around the world have been inundated with requests for placement in clinical trials. Even the online news site Psychedelic Spotlight, which usually covers subjects like drug manufacturing, has received appeals from people desperate for help. "So now . . . we have a growing demand for a therapy that is not approved, has not been demonstrated to be effective, with no actually authorized training programs for the thousands of therapists needed to meet the growing demand," reads a Psychedelic Spotlight editorial by James Kent. "This could be described as a bottleneck, but it potentially is much worse than that, it is shaping up to be a train wreck."

Headlines in mainstream news media like "Severe Depression Eased by Single Dose of Synthetic 'Magic Mushroom'" practically announce there's a silver bullet for depression. But the research it is based on is much more nuanced, and that's not something you would know unless you understand how the science is done. To get FDA approval for a drug, researchers need to prove the drug's safety and efficacy. They do this by conducting preclinical research inquiries about a drug's safety *before* it is tested on people, and clinical research studies that are done *with* people. There are three stages to a clinical study, and at every stage, the drug must succeed in order to move on to the next phase. Phase 1 studies tend to have small numbers of participants—and that is where a lot of the psychedelic research is right now. As of 2023, only one psilocybin study had made it to a large-scale phase 3 study. Most of the completed work is foundational-type studies of psilocybin that look at what parts of the brain are activated by the drug, what participants experience on it, how the drug moves through the body, the effects of different doses, and its effects on the mind and behavior. In conjunction with this foundational work are survey-type studies, animal studies, and studies composed of small numbers of carefully screened participants* that explore the efficacy, safety, and tolerability of the drug for specific applications like obsessive-compulsive disorder or tobacco addiction. They aren't applicable to the general population, but the studies can be life-changing for those who participate.

The disconnect between what the headlines suggest and what the science is actually saying might seem deflating, especially to those in search of new drugs for depression, but it takes time to get it right. Each study moves the ball down the field. But so far, psilocybin and other psychedelic drugs do look like they may constitute the first new class of mental health drugs in many years.

You don't have to count on sensational headlines to understand where the science is going. The studies are out there to be read, but a few guidelines might help you parse the serious studies from the not-so-serious. For example, solid research is published in peer-reviewed journals. That means independent experts have read and criticized the work. If you've tracked down a promising piece in a journal that does not provide peer reviewing, be skeptical. Established journals will be indexed in

* For example, the exclusion criteria for a study on PTSD at Johns Hopkins included cardiovascular conditions, pulmonary disease, diabetes, suicidal ideation, pregnancy, taking antidepressants, and dependence on a drug other than caffeine or cigarettes. And it only enrolled thirty people.

scientific databases like PubMed. Avoid journals with misspellings and formatting errors: they probably aren't what they say they are. Check out the authors' disclosures, usually at the end of the paper, to find out whom they are working for. If they are associated with a commercial interest—and many are—they may be biased toward certain results. (If you research a particular paper, you might find that some journals have paywalls, but other articles can be accessed for free. That's because in 2017, many psychedelic researchers, scholars, and practitioners signed a public statement committing to open science as a psychedelic ideal.) While the large and midsize randomized placebo-controlled study is the gold standard, a trial that is multicentered—that is, the data is compiled in mini trials from numerous sources—can be a sign the conclusions are good. And finally, look to see if the trial was registered on ClinicalTrials.gov. This will outline the goals and expectations of the study; it's essentially coming clean about their intentions.

Keep in mind that many papers on psychedelics are fueled by new imaging technologies like fMRI and PET, where radioactive tracers and MR imaging reveal metabolic or biochemical functions in the brain. (You can check the methods section of a paper to find out what techniques were utilized.) These are significant tools, but the results they produce are vulnerable to subjective interpretation.

On top of all this, psilocybin trials face an array of design challenges. The absolute subjectivity of a trip makes it very hard to standardize and measure anything. For example, if almost everyone who takes an antacid experiences relief from their acid indigestion, we can agree antacids abort indigestion. But if you trip to deal with depression, measuring results depends on how you define success, and that can vary from person to person.

Researching psychedelics in the traditional way is problematic, too. In a blinded placebo-controlled clinical trial on a cancer drug, for example, if the person who takes the drug has a reduction in tumor size and the person who took the placebo did not, the results are clear. With a dose of 30 milligrams of synthetic psilocybin, however, the subject will likely know she is tripping, which invalidates the placebo. Knowing you tripped also means that in the follow-up interviews, patients unblind themselves, making it harder for researchers to assess the value of the treatment.

The scientific community is also grappling with the question of whether psychedelic researchers should use psychedelics themselves.* Marc Aixalà, a Spanish psychotherapist and author of the book *Psychedelic Integration: Psychotherapy for Non-Ordinary States of Consciousness*, has said it depends on the context. At a psilocybin retreat or private setting, a guide probably needs to understand how very vulnerable the tripper is. "How easy is it for *you* to be held?" he asked. In a clinical trial, however, it is probably not necessary for the researchers to be experienced—how a patient responds to the drug is determined by a survey. (It's the same as asking if a nutritionist has to eat well to do their job.) But it is a complicated question, because while a scientist's objectivity could be compromised by their own use, leading to a possible bias for or against the drug, an intimate knowledge of the drug might give a psychologist insight that leads to more informed and creative studies and research questions.

Just as it is hard to find an unbiased jury to sit for a well-publicized crime, patient populations who have heard the hype

* Mazatec healers are sometimes the only ones who consume the mushroom in a traditional ceremony.

can be biased toward a good outcome. Maybe they experience anticipation of a cure that in turn provokes a placebo effect. That seems even likelier given the mysterious nature of psilocybin's efficacy; what's more, in clinical settings, the drug is often served in a vessel with great gravitas. It's a setup for success. "The placebo effect," wrote the placebo researcher Ted Kaptchuk, "is a health improvement initiated from rituals, symbols and behaviors involved with healing."

It can also lead to disappointment. In one study, a patient with depression who received a placebo fell into despair when the treatment didn't yield the hoped-for benefits. That patient attempted suicide. Accommodating the problem of expectation bias is evolving: going forward, study participants who receive a placebo may be informed afterward and given the opportunity to try the drug for real after the study is completed. One possible alternative to a placebo is to replace it with another psychedelic, like ketamine, which would ensure that patients at least receive a psychoactive drug. Additionally, psychological support in conjunction with a psychedelic trip seems to be key to the drug's efficacy. But testing this is potentially unethical because if you are going to compare the efficacy of psilocybin with and without psychological support, you risk harming the participants who don't get the support, possibly even worsening their condition.

Getting into a Clinical Trial

People are clamoring to get into studies, but because the studies are still small, opportunities to participate are quite limited. Researchers screen carefully for mental disorders that may be exacerbated by the drug, and comorbidities—even typical ones like high blood pressure—are grounds for exclusion. "Generally, if you can't engage in moderate exercise, you shouldn't be in a trial," said Peter Hendricks in an interview. "Psilocybin is pretty darn safe. But nothing is without risk." In many cases, the participants find the study—not the other way around. They are folks who have read about psychedelic therapy and have the time and resources to participate (many studies don't pay).

But because of this, diversity is a problem. Most participants are white, educated, and from wealthy industrialized nations. Additionally, Black communities remember the excesses of the CIA's midcentury MK-Ultra program, which experimented with high-dose LSD on nonconsenting Black inmates, said Kwasi Adusei, a psychiatric mental health nurse practitioner, in a talk at the first BIPOC Psychedelic Conference in 2023. Likewise, in the early years of psychedelic research, gay people endured horrific conversion therapies using LSD. If they are reluctant to participate in trials, both communities have good reason. Taking ownership of this destructive history may help researchers expand the diversity of participants.

If you want to join a clinical trial for psilocybin, you should already have a diagnosis for the condition that the drug is being tested for, like OCD, anorexia nervosa, or treatment-resistant depression. To find a clinical trial, go to ClinicalTrials.gov and search "recruiting" and "not yet recruiting" studies, then enter your condition (let's say OCD) and Psilocybin under "other terms." The studies will pop up. Check the requirements and make sure you meet them. Next, apply. If the researchers contact you, there will be additional screening, and if all goes well, you will have to provide consent, though you are not obligated to go through with it if you change your mind.

Study designs vary, of course, but if you are accepted into a trial, you will receive synthetic psilocybin or one of the patented psilocybin analog drugs. Most likely, you will have conversations with the researchers about your expectations and trepidations before your trip, which will take place at a research institution. The standard is to set you up on a bed or couch, in an eye mask and wearing headphones with a music playlist, to turn your trip inward. They will probably check on you frequently. After your trip, you will likely be asked to fill out a survey about your experience and maybe engage in a follow-up conversation. And remember, you may end up in the placebo group, if there is one. To avoid disappointment, maybe it's best to see this as helping the science, which we all will benefit from someday.

Many scientists I talked with acknowledged there could be a backlash to all the hype about magic mushrooms once the public realizes how little is known about psilocybin's efficacy, and when, as the studies widen their demographics, adverse reactions turn up. What determines an adverse reaction, however, is as subjective as everything else when it comes to psychedelics. People who report adverse reactions like having a bad trip often also report that the bad trip was meaningful to them. For example, Johns Hopkins conducted a survey of about two thousand people who reported bad trips on psilocybin. Thirty-nine percent rated the experience among the top five most challenging experiences of their lives. Eleven percent put themselves at risk of physical harm, and three people tried to kill themselves. Yet despite these difficulties, 84 percent also reported the experience was beneficial.

If public confidence in psychedelics is to endure, people will need to believe the risks haven't been covered up. Transparency is key to achieving that. Likewise, the potential benefits of psychedelic therapy need more clarification, too. *Psychedelic healing* is a misnomer if we think of it in terms of a cure. Psychedelics are more catalysts for personal and spiritual growth, pointed out William Richards in his book *Sacred Knowledge*—growth that can lead to improved lives.

Psychedelic Retreats

Retreats are group getaways, and they've been employed over history for study or contemplation free of the rumpus of everyday life. A psychedelic retreat is a modern incarnation of this: a place and time when participants join others to explore their psyches with drugs. Many retreats try to build communities by providing opportunities for participants to reconnect after the retreat is over, and all of them are competing for clients.

That is why, I think, most retreats have websites with photos of beautiful people looking blissful in gorgeous surroundings and offer gushing testimonies about life-changing experiences. They certainly look good on your phone. Retreats tend to be conducted at rented private properties in places like the Netherlands, Jamaica, Brazil, and Costa Rica, where the mushroom (or truffle in the Netherlands) is, at this writing, legal.

For years, Americans have gone to Amsterdam to enjoy magic mushrooms legally. In the 1970s, the Dutch determined there were risky drugs and not-so-risky drugs and decriminalized those in the not-so-risky category. So, cannabis made it into not-so-risky, and heroin stayed in the risky category. That led to the establishment of "coffee shops," where customers could buy small amounts of cannabis. In 1993, the first "smart shop" was opened, where legal "smart" or designer drugs were sold. Smart drugs are synthetics, like ecstasy, and over the next few years, a game of whack-a-drug began, where underground chemists developed a new drug, which came on the market legally until the government banned it, and then another drug would appear. The smart shops also purchased magic mushrooms from small growers, and eventually industrial mushroom growers got into the business of magics, but the Dutch government banned the mushrooms in 2008. However, in what may have been a mycological oversight, they didn't ban magic truffles, the sclerotia of three species of psychoactive *Psilocybe*. (Not all species produce this mycelium-rich resting body of the fungus—which is not, by the way, a truffle.) Magic truffles are generally less potent than mushrooms. The psychoactive chemicals in sclerotia are the same as the psychoactive chemicals in mushrooms, but the amount per dried gram is less and they are shorter acting. If you go to a retreat in the Netherlands, you will trip on magic truffles.

If you are curious about going on a retreat, there are a zillion options on the internet, or so it seemed to me. Luckily, I didn't have to sort through them all, because I came across Blue Portal, a facility run by Tradd Cotter, a colleague of mine from the mycological world. Tradd is an accomplished mycologist from South Carolina with a long interest in medicinal mycology, like training mycelium to produce highly specialized antibiotics. It had been a few years since we had last spoken, and I wasn't aware he'd waded into the psilocybin retreat business. He started Blue Portal with Irene Dubin, a pharmacist and psychotherapist, in 2021, moving their operation between Jamaica, Cancun, and Costa Rica, where they now have a permanent facility.

My visit to Blue Portal took place in May (coincidentally Mental Health Awareness Month). It was an adventure that I hoped would produce a new and improved me. But most of the people I tripped with were dealing with far graver stuff, like trauma, OCD, depression, and anxiety, and for them, the retreat option wasn't a lark. It was an imperative.

Blue Portal, like other retreats, offers a stay for one trip or a longer stay for two or more trips. I signed up for a four-day retreat that included two trips as well as private and group therapy. The first step was filling out an intake form and scheduling an interview with Irene, who oversees everyone's treatments. The intake form helps the organizers determine whether it is safe for you to consume the mushrooms, and if your expectations meet the possibilities of the experience. It asked about my goals, my intentions, and which specific problem I would like to address. They wanted to know my psychoactive experience—everything I've done, from heroin to beer pong—my concerns regarding the retreat, and, my favorite question, "What is your general level of comfort with altered states of consciousness?" They asked about spiritual practices, including yoga and non-dual contemplation, a kind of we-are-all-oneness frame of mind, and about my childhood environment and if I had suffered neglect. There were questions about mental health conditions, if I have a first- or second-degree relative with schizophrenia or any other psychotic disorder, and what prescription medications I take. Since SSRIs can weaken the effects of psilocybin—which, ironically, are the very drugs many folks who want to try psilocybin are or have been on—most retreats require clients go off their meds a few weeks before attending. (That might not be enough: a study in 2023 suggested the dampening effect can last as long as three months.)

In a subsequent Zoom meeting with Irene, I found her to be calming, a big blossom of a blonde in a tie-dyed muumuu, adept at the language of therapy, and focused on understanding the story of my own insecurities—namely, where I fit into the world now that I am practically a senior citizen. She talked with me for an hour, about three times as long as a typical appointment with my GP.

The Blue Portal sessions were held at a B&B called the Wharf House on the shores of Montego Bay in Jamaica. I was met at the airport by a young Jamaican man, Doug, who I figured was with the retreat because his hair was dyed blue. Already in the car was Meg, a tiny, exuberant therapist from Long Island. As we zipped along the bay road, she told me she treated patients with anorexia nervosa, and since AN might be treatable with psilocybin, she wanted to see for herself what psychedelic therapy would be like.

The Wharf House sits on a bit of sandy beach, shaded by mature coconut and banyan, ficus, and papaya trees, insulated from the waterfront traffic by coral block walls. The house is British colonial style, with dark mahogany floors and couches upholstered in colorful batik, facing fluttering table umbrellas on the patio, and, beyond that, the breezy blue bay. It's unpretentious, and utterly lovely, with the dumpy elegance of an English aristocrat's estate. There is a shingled studio structure on the premises, the space where the group conducts its trips and therapy sessions. Lots of retreats by different organizers are held at the Wharf House. It's a protected little compound with good vibes, well-suited to containing wayward trippers.

We were a group of four men and women from all over the country, plus a fifth who showed up for our second trip. That first night, we mingled awkwardly on the patio, nibbling on beef and vegetable patties, polite but reticent. That isn't always the case; I've heard reports of fellow participants looking to bond, to unload their life stories, like a chatty passenger seated next to you on an airplane. But except for my early interactions with Meg, the group was mostly shy and restrained.

Though some retreats will target specific complaints, like grief, most take all comers, and in any given group, there may be people who are dealing with an eating disorder, or OCD, depression, or PTSD, or they may be dealing with an acute crisis like a death in the family, war, divorce, or a health problem. Mixed in may be someone with no acute problems at all, just a desire to expand their consciousness. I asked Irene if people with different issues mix well. "We share enough to be comfortable," she said, a vague answer but one that suggests the dynamics of each group are unique and so defy a prescriptive approach. But in the group integration part of the trip, she said, "We share everything, and everyone learns from your experience."

Navigating the Retreat Options

Some retreats actively market to people with preexisting mental or other health issues, from depression and PTSD to chronic Lyme disease. Indeed, in 2022, the Jamaica Promotions Corporation stated that the local tourism industry is poised for further developments in the use of psychedelics for mental illnesses. You may encounter Mexican retreats, but they are considered risky. Magic mushrooms are illegal in Mexico except when used in traditional spiritual practices or ceremonies.

There are US options. In 2020, Oregon passed the Psilocybin Services Act, which was the first law in the US to establish a regulatory framework for receiving psychedelic mushrooms. In preparation, the state had to issue licenses for cultivators, labs to test the product, facilitator training programs, and service centers where the drug is consumed. The first psilocybin service center received its license to operate in 2023, and quickly thereafter, numerous other service centers opened, and some have closed. I imagine they will continue to do so until the market stabilizes and the actual number of folks who are interested in psychedelic therapy levels off. It is not a dispensary model—you must trip in the service center under supervision, and statewide, the waiting lists are thousands long. Information about Oregon's psilocybin service centers can be found on the Oregon.gov website, under Oregon Psilocybin Services. In time, other states may follow Oregon's lead.

Psychedelic retreats are not cheap. At the Eugene Psychedelic Integrative Center in Oregon, a "level 1" microdose session lasting one to two hours costs $500, a "level 2" microdose costs $800, a low dose costs $1,800, and a medium dose costs $2,800. A high-dose session costs $3,500 plus the cost of the mushrooms and includes a one-hour prep session and a one-hour integration session afterward. A facilitator stays with you the entire time, to help you stay calm or get to the toilet. But these costs are beyond many people's ability to pay, and are not covered by insurance.

Overseas retreats are expensive, too: up to $1,000 a day, excluding airfare, though some retreats offer discounts or grants. Blue Portal, for example, gives a low-income ticket, sort of like a scholarship, to those who qualify. Psilocybin-focused retreats may provide one to three trips, sometimes with individual or group therapy before and after the trip. It's hard in advance to determine what number of trips is going to be right for you, but Irene told me someone with OCD might need a few sessions to break the feedback loop of intrusive thoughts and responsive behaviors and rewire new pathways. For others, one dose may be adequate. "Just as a single calamitous incident can have lasting, crippling impact," observed James Fadiman, "so a single intense propitious experience can have lasting therapeutic effects."

For veterans suffering from PTSD, groups like Heroic Hearts, VETS, and Veterans of War provide access to vet-centered retreats and financial grants to psychedelic-assisted therapy in countries where it is legal. Heroic Hearts, for example, offers therapy for vets by vets who have been through a psychedelic-coaching program. For psilocybin work, Heroic Hearts may send a vet to a commercial retreat in Jamaica, but always with at least one Heroic Hearts facilitator, "the person that watches their back," said Jesse Gould, the organization's serious yet tender founder. Or they may completely outfit a retreat with their own people. In the long run, psychedelic therapy can mean savings for taxpayers. Care for a vet with PTSD costs about $1.4 million over the course of their life. A modest approximation estimates some six hundred thousand vets have PTSD today. So even if Heroic Hearts, which charges around $4,000 for a session, needs to send a vet back a few more times, the potential cost is so much less, like $840 million versus $7.2 billion. And if it is truly helping vets as Gould says it is, then exploring the psilocybin therapy option is not only fiscally sensible, it's a moral imperative.

But most everyone else will have to go it on their own. And there are a lot of retreats to choose from. It doesn't help that clear information regarding what drugs are offered is often buried in metonymy. "Plant medicine" is a catchall for any number of drugs, including the stimulant hapé (also known as rapé), a tobacco-and-herb snuff, and Sananga eye drops, an extract of a plant from the genus *Tabernaemontana* that indigenous peoples of the Amazon use to increase visual acuity and energy. Cannabis is the plant medicine you will find at Dimensions, a Canadian business that caters to people who "don't want to go to Peru and put a yoga mat on a dirt floor."

Some retreats are run by long-established psychedelic organizations, like the Beckley Foundation's Beckley Retreats in Jamaica. Some retreats are under the aegis of publicly traded companies, but many others are private operations, and they run the gamut from glamorous self-help spas to small operations like Cardea, which also runs a ketamine clinic in New York City and describes its style as "radical hospitality"; from Jamaica Grief Retreats, which caters to those suffering from the loss of a loved one, to BIPOC-leaning businesses like Temple of Zion Life, run by Stephanie Barnwell, an indigenous Gullah Geechee from the Sea Islands of South Carolina. It would be impossible to share a comprehensive list, as these operations tend to come and go, to close, rename, and reopen. Though you could check retreat.guru. It's like Yelp for mind expansion.

Without the oversight of a group like Heroic Hearts, people who want to try a retreat experience must figure out for themselves the legitimacy, the professionalism, and the appropriateness of the venue. A good place to start, suggested Mark Haberstroh, a guide with extensive experience working at many different retreats, is to contact the retreat founders "and ask them, How much prep work and aftercare is included in the offering? And how many retreats have you run with the same staff that will be there for this one?" Most retreat organizers rent space, but if the owners of the space are around, it's also good to determine what their involvement with you will be. Some owners of properties or organizers of retreats may see the program as an opportunity to find new converts to their religious sect, as has happened at least once in the past. Trippers are vulnerable and impressionable, as Haberstroh put it, and "No one with an agenda should have access to you during that time."

If you are going for therapeutic purposes, it's a good idea to understand the wraparound therapy package, too. According to Haberstroh, 30 percent of the experience is intake therapy or preparation, 10 percent is the actual psychedelic trip, and the remaining 60 percent of the experience, if it is to be successful, is integration and related follow-up therapy—as much as the patient/client needs. But that's more follow-up than many retreats are prepared to provide. Haberstroh explained to me that retreats must move a certain number of folks through their doors to stay in business. "When a lot of people are coming through, it is difficult to keep up with everyone," he said with a sigh.

On my first evening at Blue Portal, we met as a group in the yoga studio and settled on yoga mats in front of a makeshift shrine composed of mushroom figurines and shells from the little beach, books about mushrooms and tripping, and flowers. The altar in psychedelic use, which not all retreats employ, is born of the Mazatec tradition, where a raised table adorned with incense, candles, and images of the saints is a central focus for prayer. In our case, the shrine was, I think, a metaphor for communion with the mushroom and whatever else we wanted it to be; Irene encouraged us to add something personal of our own. I added a hermit crab shell I had found on the beach, which struck me as a metaphor: maybe I would shed a shell that didn't fit anymore, too.

Trip Guides (Want to Be One?)

Judging a retreat based on its facilities is maybe less important than judging it based on its facilitators and therapists. A lumpy mattress isn't going to cause harm, but an insensitive guide or therapist might.

Besides domestic workers and the occasional nurse or MD, retreats usually employ a combination of therapists and trip guides (or "facilitators" or "sitters") of diverse genders and, to a lesser degree, diverse races, who watch over you while you are tripping to make sure you are comfortable and don't do something stupid like try to swim across the bay. At Blue Portal, we had a total of four guides. Irene, in a flowing kaftan; Tradd, a witty and warm man who had pulled a muscle in his neck a few days before and was wearing a neck brace; Doug, whose recent mushroom experience helped him unload "a whole load of hurt"; and Pasha, a gentle Russian yogi, lithe and smooth as a cougar. Four guides, four (and then, on the subsequent trip, five) clients.

When you are researching a retreat, it is important to learn the ratio of facilitators to trippers. In Mark Haberstroh's opinion, one facilitator should not handle more than four people. A facilitator has to keep everyone under their watch safe, and if there are too many trippers for one person to keep track of, some may wander off and get in trouble. It's also informative to find out how many retreats a facilitator does in a month, "as people burn out." Facilitators may be underpaid, overworked, and isolated from their families for weeks at a time. It's not an easy job. What would you do if someone under your care started to strip? Their clients may yell, or sob, or chatter and puke, or dash about. And a session can go on for hours. The work, said Haberstroh, is emotionally exhausting.

Guides come in many shapes and sizes. Some call themselves healers or shamans; some are trained therapists; others do therapy, but their training is nontraditional. A trained therapist may be important for post-trip integration and ongoing therapeutic support, but for the trip itself, you only need a guide, said Rebecca Martinez, author of the book *Whole Medicine*, an ethical manual for trip facilitators. Martinez is the director of the Alma Institute, one of twelve psilocybin facilitator training programs approved by the state of Oregon. She pointed out that being a therapist does not necessarily make you a good trip sitter. "It's a very distinct skill set," she said. A BIPOC tripper, for example, might feel safer with a facilitator who understands their social context, including the effects of racism, intolerance, and the war on drugs. A trans person may feel excluded if a guide brings up the "divine masculine" or the "divine feminine." Indeed, if a guide tries to sell the notion that psilocybin will offer the feeling that we are all one, how does that resonate with people who have only experienced marginalization? The best professional trip sitters are aware of how profoundly social contexts can affect trippers.

I've met several guides whose own positive or healing psychedelic experience motivated them to help other people. Corinne Crone (her real name), a fifty-year-old woman with an undergrad degree in neuropsychology and philosophy, is one such guide. She grew up in a faith-healing cult. By the age of eighteen, she decided "everything I was taught was a lie, so I had to figure out what is true." In her late forties, she experienced a guided trip that in turn inspired her to become a guide. She got an advanced degree in integral psychology from the California Institute of Integral Studies and is on track to become a licensed marriage and family therapist. "I would struggle with feeling legit without a foundational education," she told me.

The benefit of a guide with a therapy or social work background is their training in ethics and transference—and they have a license that can be revoked if they mistreat a patient. The Alma Institute is very new, but it does represent a model of what skills regulators in Oregon think psychedelic guides should have. Its program focuses on ethics, interpersonal skills, transference and countertransference, and responsibility that considers the client's vulnerable state. It also requires the facilitator do their own "inner work" and become cognizant enough to stop themselves if, during someone's trip or integration, they feel compelled to introduce any of their own beliefs or worldviews, their religion or cosmology. As Martinez said, quoting one of the institute's advisers, "When I am at my best as a guide, I am a hollow bone."

But in general, it is hard to judge a guide's competency based on their training credentials, all of which are self-accredited, as there are no overarching accreditation organizations like those that exist in the medical field. In some cases, students may receive certification as guides after putting in fewer hours than you need to get a license to cut hair. Several psychedelic companies are developing or offer training courses, like Compass Pathways, as are nonprofits like MAPS. Some are training therapists to deliver specific drug protocols, and eventually those therapists will be pipelined into company-owned clinics. When researching a guide, their bios will likely list a number of certificates in disciplines, like Holotropic Breathwork, and from diverse programs, like Grof Legacy Training, developed by the eminent researcher Stanislav Grof, to marketing-savvy companies like Third Wave, where coaches learn the "Third Wave coaching methodology" (for $14,000), from the Icahn School of Medicine at Mount Sinai's Center for Psychedelic Psychotherapy and Trauma Research to the Psychedelic Sitters School. This is a worldwide industry; there are psych-assisted therapy programs coming out of Australia, Canada, Israel, and Europe, like the MIND Foundation, a nonprofit that trains medical doctors, psychotherapists, and other mental health professionals in psychedelic coaching and therapy.

I took a basic online training called the Science of Psychedelics offered by Microdose, a content provider for psychedelic education. The series of lectures by Erica Zelfand, an integrative and functional medicine physician, costs $449; more if I wanted a certificate. Throughout, pop-ups announced someone else had just enrolled in the course, giving the program a QVC vibe. It was informative and clear, except when it wasn't, because, well, the black box that is the brain. The trip-sitting advice reflected Dr. Zelfand's own work with the Zendo Project and its emphasis on harm reduction—in other words, keeping the tripper safe without getting in their way. (I also noticed the dosing values were higher than those described by researchers at Johns Hopkins: Dr. Zelfand describes a high dose as 7 dried grams of *P. cubensis* or more, and between 2 and 5 grams as a medium dose. The takeaway here is that labeling doses based on adjectives like *high* and *low* is not as specific as actual weight measures.)

One of the leading trip guide programs was the Center for Consciousness Medicine (CCM). I say *was*, because its reputation took a major hit when founders Aharon Grossbard and Françoise Bourzat were accused of covering up sexual misconduct. In at least one case, a CCM therapist allegedly tried to persuade a client who objected to being fondled that his past trauma was causing him to resist the therapist's methods, which the therapist insisted were healing. (Beware the all-knowing practitioner who thinks they are incapable of doing harm.)

Even after a trip, abuse may continue. Patients who have named their abusers have been called crazy, a negative affirmation of fears they may already have. There are cases of psychedelic therapists indoctrinating the tripper into losing their autonomy and becoming the guide's lover, student, or mentee, in some cases recruiting them to do chores like mow the guide's grass or babysit his kids.

Red Flags

Unfortunately, a guide may have great-sounding credentials but still be predatory, as was one who oversaw a trip Mark Haberstroh took. "I think I learned something about what it is like to be a woman in the world," Haberstroh told me, because "I had this guy look at me like a piece of meat for five hours. That wasn't right." He is still upset about this violation, years later.

Most of the harm that trippers have reported is sexual, where therapists caress their patients during their trips or conduct sexual relationships outside sessions. In other cases, a therapist might try to separate a patient from their bank account. George Sarlo, a survivor of the Holocaust, battled depression his whole life. He experimented with psychedelics under the guidance of a trained psychotherapist, Vicky Dulai, who received over $1 million and a Porsche, which she characterized as gifts. Sarlo's daughters, from whom he was estranged, sued Dulai for elder abuse. (The case was settled and Dulai gave the Porsche back.)

Because of the power imbalance between patient and therapist (or guide), and because of the tripper's vulnerability—even after the drug has worn off—there is no way a patient can consent to sex. Some folks will say the harm done by a psychedelic therapist is no greater than that caused by any doctor or therapist, but at least a doctor or therapist has a license to lose. Additionally, all manner of techniques may be employed in a session. If you participate in touch therapy, for example, the line between a supportive hug and feeling like you've been felt up may be blurry.

Touch is not necessarily a part of psychedelic-assisted therapy or trip facilitation. If it is unwelcomed by you and someone says it is part of the therapy, they are wrong. It is also unethical for a guide to push religious views or ask for money or accept gifts during a psychedelic experience or immediately after. Nor should they, in my view, give supplement advice, or endeavor to explain a client's state of mind with new age sops, like blaming your distress on Mercury being in retrograde. Trippers are often not in the best position to advocate for themselves—it's like being in a yoga pose you can't get out of so easily and finding yourself pawed by a yoga instructor who insists he's adjusting you—so the burden is on the guides and their employers to create a secure environment. It's not uncommon—and eminently sensible—to establish those guidelines beforehand. Generally, the job of trip sitters is to just listen, to remind the tripper she is safe, and to avoid directive therapy.

But right now, there is no ethics board that you can go to if you've been harmed, and psychedelic therapists have no governing body, though the American Psychedelic Practitioners Association (APPA) is the closest the industry has come so far. APPA has issued professional practice guidelines for psychedelic-assisted therapy in an effort to self-govern, but it's hardly perfect.

"At the moment, there is no gold-standard psychedelic therapy," Ben Sessa told me. "But one day, we will have an international global school for psychedelic therapists."

In our initial group preparation session at Blue Portal, we introduced ourselves but no diagnoses were announced. I asked for and received everyone's permission to write about my experience with the understanding they may play a part in my book, if anonymously. Tradd explained the run of show. There would be no alcohol offered at any time during the retreat. The next morning, trip day, we were discouraged from drinking coffee (as an early riser, I was, nonetheless, able to find the staff coffee machine). We would eat only fruit to minimize nausea. By 10:00 a.m., we would gather in the yoga studio for our trips. We were to stay as long as possible—at least four hours—in the studio, on our mats, wearing an eye mask. Wearing the mask keeps the tripper looking inward, Irene told us, warning, "If you take your eye mask off, you might feel the disorientation and terror of being in two worlds." That was okay because it seemed like no one was there to see the walls melt. We were there to get unstuck. Blue Portal provided iPods and headphones with different instrumental playlists to choose from, as Tradd said it's a drag to have to fuss with your phone during a trip.

Unlike most clinical trials and Mazatec ceremonies, in retreat settings, you will likely trip with a group of strangers, either in one room or situated around the facility (though there are one-on-one retreats, like Shamanic Space, which offers expensive private psychedelic sessions in the Netherlands). Most of the people I've talked with who planned to go to retreats were uptight about the group aspect of the trips. "It's like those communal tables at restaurants," one friend told me. "You never know who you are going to get . . . and if they will ever shut up."

But it can be a good experience, too. "I was a little fearful that I would try to protect myself and not 'let go' amongst other people," said a sixty-year-old artist from New York City. "But *everyone* was vulnerable, and I came to realize that what we're exposing isn't unique. Individual, yes, and needing full respect for that individuality, but not unique. So as a result, the sharing can be a benefit rather than a loss of self or specialness." Group therapy makes sense to vets, too. "Veterans are so used to being in a group dynamic," said Jesse Gould. "For them, it's a relief to not do it by themselves."

During our trip, we were not allowed to leave the studio without a guide, and no swimming was permitted until the next day. We were offered water and fruits, and if we had to pee or needed something, we could just raise a hand, and someone would be there to help. It all sounded very safe, but up until the last minute, I was working out how I was going to escape the whole adventure. "Some people are anxious or nervous," said Irene when I mentioned my initial fears. "But that's okay. It is my job to normalize it."

Throughout the evening preceding our first trip, Tradd and Irene met with us privately to discuss intentions and dosage. I didn't realize that you don't have to set an intention; I felt an expectation that I should, but I was ashamed to share my concerns about aging. At the time, my mother was in an advanced state of dementia and all but comatose; whenever I told someone, they always asked how old she was, as if her age was the indicator of how sorry they should feel about her circumstances. I felt like I was too young to complain about getting old.

When you discuss your intentions, the facilitators and/or therapists will advise you on the dosage to start with, based on your conversation about experience, hopes, and concerns. I settled on 3 grams—pretty much the same prescription as a clinical trial. I was also told a 1.5-gram booster would be available if I wanted it. In nonclinical settings, like retreats or private psychedelic therapy sessions, boosters are almost always offered to those who think they want to "go deeper." But it is easy for people tripping without supervision to take a booster too soon. You won't know within the first thirty minutes how deep into a trip you might go. Trippers who boost too soon might end up tripping harder than they expected or than their setting safely allows, potentially leading to anxiety and fear. It is also unwise to take booster doses too far along in a trip, as it could extend the experience beyond the typical four to six hours, which might, depending on the setting, have its own risks, besides being potentially exhausting.

When we assembled in the yoga studio in the morning, our doses were ready for us, the mushrooms ground and steeped in little tea bowls sitting on cards with our names. Some retreats may offer the whole mushroom, or a capsule filled with ground mushroom, or a tea. Most likely, it will be a moderately potent *P. cubensis* cultivar like Golden Teacher, which is what I took at Blue Portal, though if an experienced cultivator is associated with the retreat, other, possibly stronger species may be available. Tradd had some of those, too. We drank our mushroom tea, put on our headphones—I chose Bach's *Goldberg Variations*—and lay down. During the first hour of the trip, the facilitators pretty much everyone but Irene—played little tinkly cymbals and chimes and rain sticks and ambient music; they burned incense, came around to adjust our blankets, and then, after an hour or so, they dialed it down because everyone was hearing their own music. I declined the bump in dose and pulled out my earphones: the variations had begun to sound jumbled, the notes randomly tumbling around. I stayed still as a little mouse under my blanket, eye mask on, scanning my body with my mind, periodically dropping off into dreams.

Expect the Unexpected

Retreat testimonial pages are awash with amazing experiences. "My anxiety, a '10' going in, varies from zero to one or two." "That negative critic inside my head has just almost totally shut up." These sound like very satisfied customers, but more often, a psilocybin trip doesn't cause a complete turnabout on your troubles so much as provide valuable insights and metaphors. At the other end of the spectrum are stories—and these aren't reported as often—of anxiety-fueled experiences that leave the tripper spent and exhausted. Instead of finding themselves in an improved frame of mind, they may feel worse.

It's possible you won't get any of the answers you were looking for, and nothing of psychological value happens. Or you might get some answers, but to different questions. That's what happened to one man, Ryan, a middle-aged music producer, who went on retreat to deal with depression resulting from a divorce. Unexpectedly, he learned his junk food bingeing and subsequent weight gain since his divorce grew out of a traumatic childhood event that he watched during his trip, an audience to his own experience. When Ryan was three, he and his older brother went to day care, where they were separated by age groups. "I didn't love it and felt anxious," he said. When he saw his brother at recess, he was ecstatic. But when recess was over, and the boys were separated again, "I lost my shit." He was so distraught about being separated from his brother a second time, he was sent home, and thereafter, his brother continued at day care alone, while Ryan spent the days with his grandfather. In integration therapy after the trip, he remembered that his grandfather, who had no idea what to do with the child, ended up taking Ryan to Dunkin' Donuts every day. "That was the seed of my comfort eating," he said. He'd never connected the dots before, never realized that when he was bingeing, it was in response to the pain of separation.

I also didn't get what I asked for. I had quite a few insights about my marriage and my relationships with my adult children, but most notably, I spent part of my trip in the body of my severely demented, near-paralyzed mother. That may sound horrifying, but it wasn't. It was . . . *observational.* I don't remember how I got into my mother's body, but once there, I couldn't move. I felt like my only functioning nerves ran down my spine, as if the living me was just a skull and a snaky spine, like those creepy models you see in the chiropractor's office. All I could sense of my arms and legs was the space they were taking up. A couple of times, I wiggled my fingers to be sure I still could, and I could, so, confident I was still me, I sank back into her. At one point, I had to pee, and I worried about getting to the bathroom since I was paralyzed, then I realized I was wearing a diaper, so I didn't have to get myself up. Although I didn't ultimately pee myself, knowing I/my mother could was a tremendous relief. In fact, if what I felt was what she felt, she was mainly relieved she no longer had to do anything for herself. It didn't really matter that she could hardly move because she couldn't feel her body anyway.

At the time, this gave me some comfort, as I came away thinking she was not anguished or locked, fully cognizant, inside an unresponsive body as I'd often imagined. Instead, I think my brain accessed the significant amount of data I had acquired from years of caring for her, communicating with her doctors, and witnessing her decline. I applied that knowledge to my own body in a way that totally integrated what I knew about hers. I was in a state of extreme empathy, empathy made manifest.

And of course, relative adventures were happening in the minds of everyone in the room. The amount of work it took to keep our small group calm was surprising. There was always someone crying, someone crawling away, someone who needed to talk, to vomit, to be comforted or redirected. And most of us came out of our trips around the same time. One fellow, who took a larger dose of a species Tradd grew, *P. natalensis,* was still tripping hard, and Irene suggested we leave the yoga studio, as we were rustling about and beginning garbled chat. A meal—a wonderful country dish of chicken with ginger and scallions and rice—was laid out on the patio for us. It seemed like the most delicious thing I ever ate. Back in my room, I took a few pictures of myself, hoping to capture how elated I was feeling, but I just look a little dewy and sentimental.

I stayed relatively sober for everyone's second trip—I ate half a gram of dried *P. cubensis* and, with their permission, sat in a corner and observed. On that dose, I sat for four hours straight taking detailed notes in tidy handwriting. A fifth person joined our group, a young man whose particulars I did not know, only that he had taken his first trip a few weeks earlier. Everyone settled in, drank their tea, and adjusted their masks and headphones and blankets. For at least an hour, I watched Pasha and Doug take turns making music with their various instruments, walking around the room. Irene stayed close to an anguished man, and I saw Tradd rub her shoulders like a prizefighter's. Tradd had explained, "When you cry, we cry. We meet you where you are. Connecting with you deeply is what makes the safe container that allows you to bring stuff up to the surface. We hold you with all our energy. This is what the work is. We are connected to you individually and collectively." I don't know about holding energy, but what I witnessed was sensitive, attentive, and exhausting care. The young fellow who joined us late puked and became frightened and started crawling around. Pasha placed gym mats in front of him, laying a safe path until he eventually circled back and settled into Pasha's arms, like a very big cat, and stayed there. For hours.

Integration

If tripping plants a seed, integration therapy helps the seed grow. At Blue Portal, clients had one-on-one integration sessions with Irene between trips; they were offered art therapy if that was their thing, but I spent most of my noninteracting time floating in the bay. We all participated in group therapy and role-playing. For some, the group therapy baked in a few of the lessons they felt they'd learned. A review published in the *Journal of Psychoactive Drugs* suggests as much, finding that "social connectedness may be a fundamental, underlying mechanism of therapeutic change." But whether your integration happens in person or in a group, it's still about the same thing: making meaning of what you saw and felt. In Microdose's online psychedelic therapy class, Erica Zelfand described integration as "dry right; you've just poured cement."

After my trip, I thought I would have a session with Irene, but she said I was fine. And she was right; I didn't need to discuss my experience, and not everyone does. "Most of the time, we can do our own integration," said the psychotherapist Marc Aixalà. "We are the experts of ourselves." Integration can be as simple as journaling or walking in nature during the two-week afterglow, that therapeutic window of heightened neuroplasticity when the brain is particularly susceptible to learning. Or longer. It's not unusual for people to still be integrating a year later.

Indeed, about a year after my trip at Blue Portal, I had a new insight. Many people see their trips in different lights over time. Maybe because these experiences can be intensely vivid, the images and feelings I had didn't fade, but my perspective has changed. I still very much believe I was able to understand how my mother might have felt, because the drug made me especially empathetic. Since then, I've wondered why this vision came to me at all. I thought at first it was because I was worried about her, which is true: I was. But then I realized maybe I was also worried about myself. I've been afraid I might have inherited the genes that caused her dementia and paralysis.

I did my own integration, but some trippers need help reconciling their experience. Integrative therapy can be very useful to someone who experiences anxiety or depression following their trip or whose prior condition was made more unstable. It can help someone who was poorly prepared, or had an unresolved, difficult experience, or those who have a hard time with ego dissociation.

The thousands of people who go on magic mushroom retreats represent a significant opportunity to learn more about how different people respond, both to psilocybin but also to the variety of settings and support they receive. That's not lost on the company Maya Health, which has developed technology to help guides map patient experiences and outcomes, with the intention of sharing the information among practitioners and retreats to learn from all this nonscientific experimentation. (I met Maya's founder, David Champion, when I attended his wedding on the playa at Burning Man, along with an assortment of people in tutus, shading themselves with parasols, and a passel of folks in burlesque–meets–Lawrence of Arabia getups. It was lovely.) Maya Health is in essence a tool for wrangling citizen science.

Private Guides

A few years ago, my friend Susan lost her husband of thirty years to cancer. When I checked in on her over the years since his passing, it was clear she remained devastated. Though she had an active social life and work as an arts journalist, her grief was unremitting. Finally, she arranged a trip with a couple in Boston who procured magic mushrooms and sat with her for the duration of her journey. The experience didn't take away her grief, but it helped. "I realized," she told me, "that I can feel full and empty at the same time."

Susan's trip was a solo adventure. She knew and trusted the people who helped her. But finding and hiring a freelance guide requires the same thoroughness as checking out the credentials of a retreat. Private guides for hire run the gamut from in-person to remote, from serious to bogus (Spell and Curse removal? $220 an hour). Fees are equally wide-ranging, from the "Bergdorf-attired practitioners who bring all the things, toad, ayahuasca, mushrooms for about four thousand dollars a session," described a friend who has long interacted with people in the psychedelic scene, to an "anarchist sweetheart" he also knows, who charges $500 or even nothing; but $1,000–$2,000 a session, he said, is not unusual.

When it comes to selecting a trip guide, it's a good idea to turn your BS meter on. I listened to a podcast interview of a self-described magic mushroom healer and shamanic therapist—that's a healer, shaman, and therapist all rolled into one. Twelve times she took mushrooms. "My teachers," she said, "are the mushroom spirits."

Shaman has become a misused term. In indigenous settings, shamans like María Sabina generally emerge from and serve their community. Not everyone can be a shaman; the position is earned through knowledge, ability, or shamanic lineage, and if at any time during their apprenticeship, explained the ethnomycologist Elinoar Shavit, a shaman somehow fails, "he could be locked out." Shavit described a *curandero* she met, who was Peruvian on his father's side and Mazatec on his mother's, named Juan de Dios Garcia. "He treats sick people and cures their diseases using every method and means he knows," including psychoactive plants and fungi. That's a shaman.

Nonindigenous practitioners of shamanic arts are known as *neoshamans*, though they tend to advertise as shamans, I imagine because *neo* can sound pejorative. Neoshamans tend to be self-appointed after having worked on themselves for some time. They are in search of a community. In practice, they often (but not always) use psychedelic drugs in combination with new age spirituality. The goal is to access the spiritual realm known to indigenous shamans but lost to modernity. Neoshamans may have little to no expertise in therapy or social work, "and instead, use anecdotes and self-transformational narratives to demonstrate the power within the self to help the self, despite the obvious contradiction," wrote psychologists Patric Plesa and Rotem Petranker in their paper "Manifest Your Desires: Psychedelics and the Self-Help Industry."

Since traditional shamans usually stay close to home, people who want to participate in a ceremony end up traveling to them. I asked Mark Plotkin, a swashbuckling ethnobotanist and author of *Tales of a Shaman's Apprentice*, among other books, what defines a shaman. "I don't know how to give an accurate description," said Dr. Plotkin. "Keep in mind that there are good medical doctors and terrible medical doctors. I feel the same way about shamans and neoshamans. But people who have been to Esalen for the weekend [a holistic retreat and educational institute located in Big Sur, California] and did some mushrooms and suddenly declare they are a shaman? I have a problem with that. In fact, I once asked a fabled Amazonian medicine man how long it takes to become a shaman. He was ninety-three years old, and he said he didn't know: he was still learning!"

Most self-described shamans are rather average as a group, exhibiting "less narcissism than MBA students," said Kate O'Malley, who conducted a survey into neoshamanism when she was a clinical coordinator at the Substance Use Research Center at Columbia University. While she noted that those who were most likely to cross boundaries with their clients may have chosen not to participate, she nonetheless did see outliers who scored very high in narcissism and Machiavellianism and very low in conscientiousness—hardly desirable characteristics in a trip guide. Though limited, the survey demonstrates that neoshamans do not necessarily have Western medical training nor training in traditional shamanism. That doesn't mean they can't do the job, but without accountability, it is hard to have confidence a neoshaman will always act in the best interest of the people they serve.

Mark Haberstroh told me it's important to do your homework before engaging the services of a retreat or guide, because "there are a million ways [psychedelic treatment] can be done." Some guides will sculpt a trip using a series of psychedelics, so-called hippie flipping (administering psilocybin plus MDMA). A psychologist I know was into "Jedi flipping," alternating LSD, mushrooms, and MDMA. (We met at an event called Trip Tales, held by the Brooklyn Psychedelic Society. At one point, she asked rhetorically if somewhat wearily, "After you have plumbed your own psyche and you are done tripping, then what?")

Other guides may take mushrooms at the same time as you, maybe to increase their empathy. In a private psychedelic-assisted therapy session, a guide may employ any number of therapeutic techniques, like sexual healing, Holotropic Breathwork, ecstatic trance states, nurturing touch, Sensorimotor Psychotherapy, and Hakomi, to name a few. All of these are things I would want to know about—and have a say in—ahead of time. Red flags, for me, are guides who preach a mash-up of religious traditions from all over the globe, dressed in personal testimony. Guiding in conjunction with the wisdom of Greek oracles, druids, mother goddesses, astrology, or bees is a hard sell for me.

When appraising retreats, guides, and healing techniques, I try to use the same critical thinking I exercise when judging the claims of wrinkle creams; if it sounds dubious, I don't buy it. But on the other hand, I know that if I am feeling vulnerable, I can easily ignore my own suspicions. I think that could happen with some folks who are psychedelic curious as well. What you are buying when you go to a retreat or work with a personal guide is access to the mushrooms, set and setting, and maybe talk therapy. Many organizations and their employees are thoughtful, dedicated people. Many of the psychedelic guides who have put out a shingle are serious and well-intentioned. But keep in mind it is a business where you are the one who bears the risk.

PART III

Trip Types

8

Microdosing

Of the different trip types, all of which are dose-dependent, I think microdosing has aroused the most interest among the mushroom-curious, probably because it's said you don't have to trip to reap the benefits. Since starting work on this book, I've been asked what I know about microdosing by various relatives, a real estate manager, a hairdresser, numerous writers, an architect, and almost everybody I sit next to at a dinner party. In that same period, I found out plenty of people I know are already doing it: a socialite, a farmer, several stay-at-home moms, a chef, a doctor, various small-business owners, and a handful of aging artists. If I extrapolated from the number of people among my acquaintances who microdose mushrooms to enhance their mood and productivity, I'd end up with results like the 2021 Global Drug Survey did: of thirty-two thousand people who use psychedelics, 25 percent reported microdosing in "the last 12 months."

Who wouldn't be attracted to the idea of an enhancement drug with no known side effects? It sounded good to me. One of my professional worries these days is the state of my working memory. How much less of my research can I access—without notes, at any given time—than I could a decade ago? Working memory matters a lot when it comes to writing a book. As my husband once said, "I'd microdose to deal with aging, but I keep forgetting to take it."

I figured if microdosing could help enhance my focus and creativity as many people have claimed, I was all for it. I microdosed for a month, and it did improve my work habits, but not in the ways I expected. It seemed to relax my frame of mind, which in turn improved not just my time in front of the computer but the rest of my day as well. And that was surprising: I don't suffer from debilitative depression or anxiety, and yet something just . . . lifted. I found I could just rise above the bullshit that competes for space in my brain.

Microdosing psychedelics is the practice of ingesting tiny amounts, so small as to be subperceptual and not impair cognitive function. On the right dose, you should be able to do everything you normally do in your life. That means no hallucinations, no trance states, no ego dissolution. The term is not unique to psychedelics. In pharmacology, microdosing is used in drug development to determine how a drug moves through the body. It's a standard measure, consisting of 1 percent of a pharmacologically active dose. In the psychedelic world, microdosing is generally defined as 5–10 percent of a psychoactive dose. And therein lies the rub. What determines a psychoactive dose is utterly subjective, and that is complicated by the fact that microdosers don't usually have access to synthetic psilocybin, which can be consistently and precisely measured.

Any psilocybin-containing mushroom can be used in microdoses—though there is variation in the psychoactive content of different *Psilocybe* species. *P. cubensis* typically ranges from 0.3 to 2 percent psilocybin per dried gram, but other species can be lower or higher. Because magic mushrooms vary in strength from species to species, each person must determine their own dosage based on the mushroom species they have accessible—usually *P. cubensis*—and they commonly do this through trial and error. "It is," said the herbalist Christopher Hobbs, "a big giant experiment."

Many microdosers see themselves as conventional citizens, wrote the authors of the paper "Narrative Identity, Rationality, and Microdosing Classic Psychedelics," and that may in part explain its widespread appeal. Since a microdose doesn't make you high, and you aren't taking it to get high, you're not a druggie. Instead, many microdosers consume psychedelics for wellness, like taking a vitamin, or for productivity, like drinking coffee. One of the authors of the study, Peter Hendricks, noted this puritanical justification. "People who microdosed said, 'I realize it's illegal, but I'm a more productive worker.'"

Microdosing is becoming socially normalized, even though it is federally prohibited and illegal in most states. You wouldn't think so, based on its ubiquity in the news and social media. People are explaining how to microdose on Instagram, gesticulating with their dried mushrooms, and in YouTube videos, seminars, and summits, and consultants and microdose gurus offer dosing services all over the internet. A Google search for microdosing yielded well over eleven million hits, a significant increase since 2015; and in the spring and summer of 2023, trend data showed there were about fifty thousand searches a month, mostly asking, "What is microdosing?" The Reddit community on r/microdosing, which offers peer-to-peer advice, is over 250,000 members strong. And of course, as is the case with psychedelic use in general, celebrities are going public with how mushrooms have changed their lives. "Microdosing mushrooms," the comedian Chelsea Handler told Jimmy Fallon between swigs of a margarita on *The Tonight Show*, "is a game changer."

For decades, microdoses of LSD have been used by a quiet underground of cognoscenti that included luminaries like Albert Hofmann, though microdoses of mushrooms may be a far older practice; they are reportedly used by indigenous people in Mexico, said Elinoar Shavit, the ethnomycologist, to increase energy. But in 2011, the notion of microdosing was mainstreamed in the US with the psychologist James Fadiman's book *The Psychedelic Explorer's Guide* and the subsequent publicity it generated. Fadiman's book covers many aspects of tripping on LSD with intention, but its detailed descriptions of microdosing protocols—how much and how often—and testimonies from happy users seeped into the popular culture. Subsequent articles in *Rolling Stone* and *Wired* described the rise of microdosing for work performance among Bay Area software engineers, biologists, and mathematicians who claimed it induced the flow state, a state of optimal performance. I met someone in 2014 who was the quintessential microdoser: a San Francisco–based scientist/entrepreneur in the biotech field. He told me he microdosed LSD for company creative meetings. "I microdose because it enhances my creative flow, but suppresses my ego, which makes me a better collaborator." An appearance by Dr. Fadiman on Tim Ferriss's popular podcast didn't hurt; it introduced thousands more to the idea of a new, all-natural smart drug that even in tiny doses supposedly could lift the intellectual or emotional barriers that get in our way.

There is some science to back up these claims. In a small Dutch study, which was neither blinded nor placebo controlled, thirty-eight participants consumed microdoses of magic truffles (the sclerotia of some *Psilocybe* species) and took a series of cognitive tests. The researchers found two modes of thinking that may be key to creativity were positively affected by the drug: convergent thinking, or problem-solving, and divergent thinking, the ability to come up with multiple solutions for a problem. That's good news for someone who is trying to put together the puzzle that is a book (me), and a potential asset from a corporate perspective. The mushroom coffee company MUD\WTR, for example, practically encourages microdosing at work, because "we believe that creativity is the new productivity," wrote CEO Shane Heath. "And while high-dose psychedelic trips should be done in a safe setting (with an Alex Grey painting hanging on the wall), microdosing—that is, taking a fraction of a full dose—has a low risk profile. It can *italicize* consciousness just slightly, allowing us to make new and novel connections that aren't typically available on an average Monday morning."

I was looking for performance enhancement, but people microdose for a variety of reasons. It's possible that microdosing psychedelics is much like taking stimulants for ADHD, and already there is some evidence that suggests it may be more effective than conventional medication. And unlike Adderall or Ritalin, which can cause a range of associated problems, from racing hearts to withdrawal symptoms, low-dose psilocybin (and LSD) has virtually no known side effects. Others—many, judging from my conversations—are trying to lift their symptoms of depression and anxiety. That possibility was explored in Ayelet Waldman's book *A Really Good Day: How Microdosing Made a Mega Difference in My Mood, My Marriage, and My Life*. Unhappy with her antidepressant regimen, Waldman took 10 micrograms of LSD twice a week to relieve her symptoms.

Her glowing testimony, and others like it, has fueled hope that low-dose psilocybin or LSD can replace some prescription mental health medications. More recently, studies have shown that microdosing may indeed have a beneficial effect on mental disorders (in rats), and adults who microdose may have lower levels of anxiety and depression than non-microdosers.

So yes, I certainly wanted to try. Microdosing is individualized by the dose size—the maximum dose you can take that is subperceptual—and the protocol, the schedule on which you take the drug. The point of a protocol is to avoid developing tolerance or insensitivity to the drug, which can happen with consistent use. A friend gave me a dozen capsules containing 0.25 grams, or 250 milligrams, of dried *Psilocybe cubensis* powder. The strain was the common Golden Teacher (the "Starbucks of psychedelics," she said). *P. cubensis* in general contains about 1.25 percent psychoactive alkaloids. That's about one-tenth of a moderately high dose (for me) of 2.5 grams. My plan was to microdose every third day for a month, or about twice a week, and to take the dose first thing in the morning.

The effects, if any, of microdosing aren't an either-or proposition: either it improves mood *or* it enhances productivity. The two responses may well be complementary. For example, one microdoser told me she was more aware of her thinking and subsequently was able to arrest ruminative thought loops. Indeed, a review study in 2020 found that microdosing LSD or psilocybin had a subtly positive effect on cognitive processes, and the authors thought maybe cognitive flexibility somehow contributed to a decrease in rumination and therefore had a positive impact on depression.

The only caveat to this great news is the microdosing movement is based almost entirely on anecdote. Most of the studies on microdosing are aggregations of self-reported claims gathered on crowdsourced sites like Microdose.me or folks who were recruited from online self-reporting forums. While these real-world reports do show "increased energy, improved work effectiveness, and improved health habits . . . in clinical and non-clinical populations," to quote an observational study conducted by Dr. Fadiman and colleagues, there is a lot missing, like the participants' drug tolerance level or whether they are on medications. Microdosers also claim the drug provides a range of benefits, some of which have more anecdotal support, like migraine relief, than others, like curing shingles.*

Microdosing may be more about faith than fact. While there are numerous randomized, placebo-controlled studies on microdosing psilocybin in clinical trials, the completed studies that utilize such controls have not shown that microdosing does much more than a placebo. It's possible that any mood-related differences microdosers feel are the result of their expectation that they will feel *something*. Indeed, the main reason why people stop microdosing is that they end up feeling nothing at all.

"There's a little deflation among the scientific community," said Vince Polito over the phone. He's a senior lecturer in the School of Psychological Sciences at Macquarie University in Australia and a researcher in how altered states can affect mental health and enhance cognition. "But in my view, a pessimistic framing is premature. There are questions about how representative the study samples are of the general population." In

* Similarly, all kinds of benefits have been attributed to microdoses of *Amanita muscaria*, including improved sleep, improved mood, addiction rehabilitation, cessation of hot flashes, and resolution of gingivitis (which also may be resolved by brushing your teeth more often).

contrast to current studies where healthy people are all given the same dose, self-reporters who claim to have benefited from microdosing vary in the size and frequency of the dose they take, and the state of their mental health is undisclosed. Additionally, microdosers tend to keep it up for months, not weeks, as described in most studies. However, Apex Labs received approval from Health Canada in 2023 to begin the "world's largest take home psilocybin clinical trial" for 294 participants with depression to try microdoses of their synthetic psilocybin drug. That's good news, I think, because anyone who has taken SSRIs knows that serotonergic drugs can take a while to show improvements. I hoped a month of taking Golden Teacher would make it evident that it worked or did not work for me, though I recognized if it did work, it might be because I hoped it would.

Who's Dosing

I didn't have much at stake by microdosing for performance enhancement, not compared to those who are hoping to mitigate the symptoms of depression and anxiety. It is their testimonies that are particularly compelling. Take Larry Campbell, a member of the Canadian Senate. He'd been on antidepressants for decades when, during the pandemic, he started to feel much better. It turns out his wife was spiking his coffee with psilocybin. (While it's not my place to judge her, in general, it is unethical and ill-advised, not to mention illegal, to give anyone a drug without their consent. It's also obnoxious to give them to pets.) Today, he's a leading voice in the debate on how Canada should address psychedelic medicines.

I met a medical doctor from New York who suffers from

depression and addiction issues and has been tweaking his medication for years. He explained he is currently on a good cocktail. He got off his SSRIs and now, in addition to a minimum dose of Wellbutrin, he microdoses 10 micrograms of LSD on day one and 150 milligrams of *P. cubensis* on day two. He takes the third day off. He is not alone in replacing his medications with microdosing *Psilocybe* mushrooms. The 2021 Global Drug Survey I mentioned earlier found of those who microdosed, almost half had reduced or stopped their prescription medications. "LSD is by far the most effective antidepressant I've ever taken," he told me. "It allows me to let go of things that would otherwise upset or derail me. But the mushroom helps me open up and engage with things coming at me that are beneficial."

Moms on Mushrooms

A subgroup of the microdosing-for-depression demographic is moms. It's common for new moms to have bouts of the "baby blues," but 10–20 percent experience serious postpartum depression, where they feel empty and emotionless and unloving toward their babies. What's more, every time a mother packs her kid into the stroller, she risks social judgment about her parenting style, breastfeeding, going back to work. I remember thirty years ago taking my ten-day-old daughter out for her first breath of fresh air on a sunny July day—and my first day out, too—and being stopped more than once by people who told me I was being reckless taking such a young child outdoors. That's still going on.

More recently, I attended a sharing circle with young

mothers called Moms on Mushrooms (M.O.M.), a platform where mothers who use psychedelics can connect. (M.O.M. has garnered national attention: in 2023, the founder of the site, Tracey Tee, explained microdosing to Dr. Phil.) There were nine women on the Zoom call, from all over the country, including Hawaii. All were in their thirties and forties, caring for young and middle school–aged kids. A few had teenagers. All described the challenge of maintaining their own identities within the maelstrom that is motherhood with the help of psilocybin, sometimes by means of larger trips, but mainly through microdosing (and all gave me permission to write about the meeting). I should point out none of these moms described microdosing while they were pregnant or breastfeeding, and no studies have been conducted to determine the safety of what seems to me an uncommon practice.

One woman said she had been pregnant or breastfeeding for the last eight years. "Motherhood is a shift in identity," she said, "but it's important to maintain your sense of self rather than seek to return to yourself when the kids are grown." Since that phase of her life ended, she has been microdosing psilocybin to help rediscover and maintain her private identity—who she is besides being the mother of her children. Another mother said she microdosed psilocybin to be more present with her kids. "Microdosing is helping me slow down. I've been trying to live up to other people's expectations rather than living in the moment." Yet another described how microdosing psilocybin helped her process grief over a death in the family. A few wondered where they would be today if they had started microdosing postpartum. "How much better is that than a doctor saying, 'Here's some Zoloft. See you in fifteen years'?" For me, I kind of regret I didn't know about microdosing as a young parent. Maybe it might have helped me be less reactive and more responsive.

While the specific women in the M.O.M. group I attended all happened to be white, moms of color are also part of this trend. They deal with the stresses all mothers face, from the mundane to the existential, including raising a child in a world fraught with dangers like the threat of autocracy and environmental disaster. But Black moms who microdose do so at additional risk. Child Protective Services takes away the children of Black people more readily than those of white people, and our legal system is more punitive toward Black people than white. It makes the prospect of self-medicating with an illegal substance like psilocybin—even if it truly helps—a more dangerous choice for people of color. As one mom told the *Guardian*, "I think it's really important that black mothers do this healing work because they can find liberation, and a freedom that is not allowed to us in our day-to-day lives. It gives black women their power back. It reminded me that I'm allowed to be seen and heard."

Like many microdosers I spoke with, Chen Yu (not her real name) felt ill served by her postpartum depression medication. She said the side effects from the pharmaceuticals seemed worse than the depression itself. "I felt my emotions were blunted at both ends. No depression, but no joy. I thought I could *will* myself into happiness, but that doesn't work very well." A tough divorce was the catalyst that turned her to psychedelics. The Chen Yu I met, a TV chef and food blogger, was high energy and outgoing. "I started showing up in my life. With microdosing, I don't feel any [psychedelic effects], but at the end of the day I think, *Huh. That was a good day.* And when things do go wrong, I am able to flip them, instead of ruminating. That has been one of the biggest benefits. I started living in the present lane. Not the past lane that is rumination, or the future lane, which is wishing."

More Microdosers

Some folks use psilocybin microdoses to quit drinking or as an alternative to drinking. One friend told me, "I take such a small dose that it is really part of my almost daily routine. Much as I love my wine, it was messing with my sleep even in relatively minor amounts, and I have been a horrible sleeper for years. Now I sleep great and wake up feeling so good. Joe [her husband] kinda hates it because he is wedded to his martinis, but it's a new life for me." Even *Elle* magazine suggested that microdosing mushrooms is a feel-good, nontoxic alternative to a cocktail. And it's low-calorie! "Bye-bye, wine," reads a gleeful subhead. "It's mushroom time."

Chronic pain is another typical use. On Reddit's microdosing threads, there are testimonials from women who microdose to manage PMDD and PMS. Years ago, I met a young architect who treated his headaches with low doses of psilocybin. He learned about it on the website Clusterbusters, which had devoted a page to testimonials—both positive and negative—on psilocybin's efficacy. The anecdotal data that has accumulated on Clusterbusters has been persuasive enough to intrigue researchers and there are, as of this writing, various studies in clinical trials.

These microdosers are citizen psychopharmacists. Like so much of the research when it comes to psychedelics, citizen use is leading the science. I imagine it is just a matter of time before a researcher looks at the anecdotal data and designs a study on microdosing and alcohol consumption, or microdosing and postpartum depression.

Side Effects

What, if any, are the side effects and risks of microdosing? As with larger doses, there may be a risk of valvular heart disease with chronic use, but the risk hasn't been comprehensively assessed yet. Other, less problematic side effects depend on dose size. It's not unusual for too low a dose to be as uncomfortable as one that is too high. Why is a mystery. But one reason people stop microdosing is negative reactions. Folks have reported feeling sweaty, anxious, or headachy, or experiencing fear and frustration. On surveys, microdosers have reported insomnia, nausea, reduced appetite, impaired focus, both excessive energy like jitteriness and inadequate energy like brain fog, and social challenges like feeling awkward or oversharing. Additionally, some microdosers have reported that they cry for about an hour after ingestion. It may also exacerbate some conditions, like making a hangover more awful.

But generally, microdosing seems to be well tolerated by most people, though the reports of negative reactions are like reports of positive ones: It's hard to predict from anecdotal data what will or won't happen to you. As reported in the forums, for some people, drinking coffee while microdosing causes heart palpitations; drinking alcohol may bring on headaches or magnify feelings of depression. If you smoke cannabis, the effects of both the mushrooms and/or the weed might be stronger.

Studies have shown that taking SSRIs appears to weaken the effects of psilocybin. Whether that is true when microdosing psilocybin is unknown.* However, microdosers take mushrooms for anywhere from a week to years, and it is possible there could be an accumulative effect on serotonin receptors. I asked Emmanuelle Schindler, who studies psilocybin and headache disorders, to unpack what happens to the receptors. She explained, "If receptors are being pummeled by neurotransmitters"—and remember, psilocin functions like the neurotransmitter serotonin—"the receptors will be absorbed into the cells and not reemerge until after the drug is gone." And that may lead to decreasing efficacy of the dose. And indeed, one of the most cited effects of acute use is tolerance or insensitivity to the drug. That's why many microdosing consultants recommend protocols that call for pausing between administrations of the drug. And by the way, you can't circumvent the development of tolerance by alternating psilocybin and LSD.

Getting the Microdose

It seems the primary concern people have about microdosing is the fact that it is illegal. And they are concerned for good reason. Take the case of Jessica Thornton, an Indiana neonatal intensive care nurse who suffered from treatment-resistant depression. She microdosed *Psilocybe* mushrooms that she grew at home. They helped her cope. But someone with knowledge of her practice contacted the authorities; her home was raided, and

* Although antidepressants are one of the most widely prescribed medications in the US, according to the CDC, not much is known about their long-term efficacy and safety either.

she was charged with felony drug possession in February 2022. The arrest was based on the weight of the mushrooms she had grown plus the mycelium-infused matrix, which increased her charges. When the prosecutors understood what she was—and wasn't—doing, the charges were reduced to a single count of felony possession. Thornton accepted the plea and was sentenced to eighteen months in prison in October 2022, all suspended. She was placed on 180 days house arrest (reduced to 90 days) and ordered to attend a treatment program. As evidenced by Thornton's experience, getting the microdose presents one of the more problematic aspects of the entire endeavor.

Microdosers need a steady supply, and buying on the unregulated gray market is risky, legally as noted, but also because the chances of getting ripped off are very high. It's common for moderators of online microdosing chat rooms to ban the solicitation of drugs. You will likely be kicked out of the room if you source mushrooms or link to vendors. Likewise, the moderators say, in bold, "Do NOT reply to messages from strangers that offer you anything, because they are most likely a scammer." And indeed, it's risky to buy anything illegal over the internet. Not only is there no guarantee you will get what you paid for, but do you really want to swallow a capsule you bought from some random PO box? Legal suppliers in the Netherlands have developed microdoses of psychedelic truffles, but they won't send their products to countries where they are illegal.

Keep in mind that the wellness industry has jumped on the microdosing trend; I've seen advertisements for microdoses of mushrooms that are simply functional foods. A product like Mojo Microdose, for example, doesn't contain *Psilocybe* species at all but rather mushrooms like lion's mane and *Cordyceps*. Microdosing nonpsychoactive mushrooms is kind of like microdosing lettuce. It's moot because functional mushrooms have no perceptual effects in the first place. You can eat an enormous plate of lion's mane (they're tasty!) and what you'll feel is full.

Those who buy a whole mushroom avoid the problem of not knowing exactly what's inside a capsule they might purchase. Those who grow mushrooms avoid the problem of interacting with the illegal drug trade, though they still risk arrest. Chen Yu grows her own mushrooms because "it's always important to know where your food comes from, but I think knowing where your medicine comes from is even more important." Her mushroom bins, exploding with chubby Penis Envy, are virtuoso acts of amateur mycology. As the mushrooms flush, she collects them and freeze-dries them, which she says preserves the actives in the mushrooms better than regular dehydration. The resulting mushroom is the consistency of a Cheeto: "Way easier to eat and easier on your stomach than gnarly woody stems." She then grinds all the mushrooms together.

This is a common technique that many microdosers I spoke with recommend because there can be variations in psychoactivity between individual mushrooms of the same species. By mingling the mushrooms, she homogenizes the batch, creating a consistent psychoactive level for all the capsules she will prepare. But that's not the only way people microdose. Christopher Hobbs thinks grinding up the mushrooms undermines potency. Psilocybin and psilocin are unstable, he says, especially when exposed to light, and that instability is exacerbated by creating a powder, which has a lot of surface area. Hobbs, who told me he has been microdosing mushrooms on and off for forty years, prescribes microdoses snipped from whole mushrooms and weighed. "There are facilitators who will say up to one gram is a microdose," he told me, "but I don't see that at all. There should only be a lift in mood, a feeling of being more connected and mindful, but it doesn't interfere with your day. To me, that's what defines a microdose. It should just make you feel happier to be alive."

About Dosing

Getting to the "ideal" microdose is a subjective and individual process that can only be achieved through trial and error. On day one of my microdosing experiment, I took 250 milligrams of *P. cubensis*, Golden Teacher cultivar, and spent the first ninety minutes having hot flashes, yawning, and feeling a little nauseated and woozy, not unlike when I once had vertigo from a rogue calcium particle in my inner ear. After that, I got into a groove and spent the whole day reviewing research—a step I often find nerve-jangling, as I worry about neglecting something important. This time, though, I didn't revert to that kind of self-questioning. I just went into a flow state where I was focused and productive, and at the end of the day, a little crabby—pretty much par for the course. Days two and three, I continued to be my normal productive self, only without the queasy feeling. Day four, I took another cycle of 250 milligrams with the same initial symptoms, which told me that it wasn't just a matter of acclimation: the dose was simply too strong for me, and I'd have to start titrating.

That's pretty typical of the process. Chen Yu microdoses 200 milligrams of mushrooms three to four days a week during the months when she uses them. The challenge, she says, is a dose low enough to be undetectable but high enough to function as a medicine. That's what users call *the sweet spot*, she says, "when you feel steady and ready for the rest of the day."

But it can take time to figure out that dose. It took me ten days. On day seven (days one and four I took 250 milligrams), I dumped the contents of the cap into a little jewelry scale I bought for fifteen dollars online and measured out 100 milligrams, thinking I'd titrate up from there. I had awoken that day with a headache and had hoped it would help—as large-dose psychedelic experiences have done—but all it did was tamp down the pain a bit. My productivity was good, and the next day, free of headache, I was working well and in a cheery mood. But day eight was a difficult day, in that I had a hard time getting into the flow. My attention was disrupted and jumpy; my brain felt like a bicycle skidding down a gravel hill. I was really looking forward to microdosing again, in hopes of getting past this disjointed thinking. It didn't occur to me that maybe the bad day I had was caused by the drug.

I upped my dose to 150 milligrams on day ten. But how much psilocybin was I ingesting? I emptied and measured the weight of the size 00 gelatin caps I was taking. The empty cap weighed 130 milligrams. The filled cap weighed 410 milligrams, which means my original dose contained 280 milligrams of the *P. cubensis* cultivar Golden Teacher. Four samples of Golden Teacher mushrooms tested at the 2021 Psilocybin Cup measured 0.47 percent, 0.53 percent, 0.75 percent, and 0.84 percent tryptamines (a combination of psilocybin and psilocin) per dry weight gram. For mathematical simplicity, I chose to calculate based on the 0.75 percent strain. If 1 dry weight gram of Golden Teacher cultivar is 0.75 percent psychoactive compounds, or 7.5 milligrams, the (approximately) 250 milligram caps I had contained 1.8 milligrams psilocybin and psilocin. (The rest of the content is trace amounts of other psychoactives and dried mushroom fiber.) A high dose, according to scientists at Johns Hopkins University School of Medicine, is 20–30 milligrams of psilocybin, so a 250-milligram cap is close to the 10 percent usually described as subperceptual. But it was too high a dose for me.

I was good at 150 milligrams or about 1 milligram of psilocybin. Because I didn't have a microspoon to fill a tiny cap with powder, I poured the dose into a cup and filled it with hot water, let it steep until it was cool enough to swallow, and drank it down. I always took my dose in the morning before I ate anything. Some people add the powder to their coffee in the morning, their hot chocolate, tea, smoothie, or juice. Others make a sublingual mushroom tincture, to be used by dropper or spray, which is what LSD microdosers use. (I once stayed at a friend's house in the Bay Area and found in her medicine cabinet a bottle of Visine with a label that said "NOT VISINE!") These various drug delivery methods are one of the categories that psychopharmacological companies are exploring, even though the efficacy of microdosing is far from established. For example, in 2020, Silo Wellness, a publicly traded psychedelics company founded in 2018, filed a patent for a psilocybin metered-dosing spray intended for microdosing. Silo claims the spray system avoids nausea because it bypasses the gut. Alternatively, the biopharmaceutical company Psilera has a patent for a transdermal patch that would allow patients to dose a psilocybin-like drug called *psilacetin* topically, kind of like hormone patches.

Generally, researchers have defined a microdose as somewhere between 100 and 500 milligrams of *Psilocybe cubensis*, but there is no standard. This bears repeating: *there is no standard.* Many folks don't even know what dose they are taking. The advice I've encountered most frequently recommends starting with a first dose of 50–100 milligrams and titrating up from there. A microdosing guide from Oregon whom I'll call Tonya, a hard-knocks gal who said psilocybin helped her learn how to take care of herself, explained to me that she takes people up 50 milligrams at a time, starting with 100 milligrams, until they get to a dose they feel. Then she dials it back to the previous dose. She explained you will know a dose is too high if you experience *anything*, as in my case, drowsiness and dizziness. If the dose is too high, Tonya recommends just soldiering through it and adjusting the dose down next time your protocol calls for dosing. Same goes for a low dose: she says to wait until your protocol calls for dosing again before increasing the amount you take.

Cultivars Matter

Most microdosers use *P. cubensis*. Terence McKenna said, "A cube is a cube"—*cube* is slang for *P. cubensis*—but the people who are using these mushrooms generally say cultivars do matter. New cultivars are being developed to increase potency, so the microdoser who determines she needs 200 milligrams of a low-potency strain like B+ might find she needs to *lower* the dose if using a high-potency strain like Tidal Wave. I asked Jordan Jacobs of Tryp Labs about this. He studies psilocybin content in different mushroom species and cultivars. "Potency levels can be all over the place," he wrote me, "for both cultivated *cubensis* and foraged mushrooms."

Like mushrooms, magic truffles have varying potency, too, from 0.75 to 1.75 milligrams of psilocybin per fresh gram. According to an analysis done in 2012, the sclerotia of *P. tampanensis*, a.k.a. the Philosopher's Stone—which is rare in the wild but cultivated today—has between 0.59 and 1.67 milligrams of psilocybin per gram of fresh sclerotia. The Psychedelic Society of the Netherlands reports a recreational dose of fresh truffles is 10 grams and a microdose is 1 gram fresh, and around ⅓ gram dried.

I'm not convinced different cultivars have specific effects, but many people on the microdose (and macrodose) scene believe they do. For example, I know someone who prefers the Huautla cultivar because he finds it more of a "heart opener" than others. Microdose consultant Tonya says Golden Teacher provides "increased clarity and improves memory recall." She recommends the Lizard King cultivar for patients with anxiety and depression, and Rusty White for introspection and enhanced capacity to connect positively with your environment. People's subjective experiences are real, but not necessarily transferable. One person's increased clarity might be another person's heart opener. With all this variability, if the user is going to be precise, every go-around with a particular cultivar or species, or even a new batch if you are home growing or collecting in the wild, could require starting the process of finding that sweet spot all over again.

Protocols

The protocol I followed was one James Fadiman described in his book *The Psychedelic Explorer's Guide*. His protocol is for LSD, but many psilocybin microdosers use it. The protocol calls for one microdose day, then two days off. Dr. Fadiman found that users had an "afterglow" for forty-eight hours after microdose day; waiting two days for a subsequent dose gives the user ample time to return to one's ordinary state of consciousness. In his book, he recommends being conservative about the protocols, sticking to the amounts and days off, maintaining normal daily patterns of exercise, sleep, and meals, and discretion as to whom you let know what you are up to. After a month, he suggests, ask yourself how you feel.

Another widely used protocol is the Paul Stamets protocol: four days of microdosing followed by three days off. The well-known mycologist and entrepreneur recommends the patented Stamets Stack: a combination of 50–100 milligrams dried magic mushrooms, 200–400 milligrams dried lion's mane mycelium (for its potential neurogenerative effects), and 25–50 milligrams of the B vitamin niacin. The latter is a vasodilator that may enhance the uptake of psilocybin, which Stamets believes enhances the neurogenic benefits.* The Stamets Stack is a popular concept, as well as a brilliant bit of marketing that has resonated with other purveyors and individuals, who "stack" with all kinds of ingredients, including raw cacao powder, matcha, and ashwagandha.

While the Fadiman and Stamets protocols are the most well known, they are hardly the only ones. The Microdosing Institute, a Dutch education, research, and community platform, describes numerous protocols, including their own, which is every second day for four to eight weeks. They also describe the two-days-a-week protocol, on which users microdose for two fixed days, three days apart (say, Mondays and Thursdays). For those who get sleepy from low-dose magic mushrooms, the institute devised the nightcap protocol: every second day before

* The recommended daily dose of niacin is 14–16 milligrams, so people who use a range of medications containing niacin need to be aware of increasing their load.

bed (but not with LSD, they point out). This strikes me as the art of naming the obvious, and that it is all ultimately what the institute calls *intuitive microdosing*, or "as needed." In short, they all recommend doing your own thing, but leave at least one day between doses.

Most users take their dose in the morning, but plenty of microdosers prefer a nighttime regimen, which they say helps them sleep and feel invigorated or with improved mood the next morning and through the day. Others must time their microdoses so that they fall asleep when they want to, following a burst of energy. "It's a fine balance," wrote OutMyPsilocybin on the Reddit forum r/microdosing. I recently met a man who splits a daily dose into two, one in the morning, the other at night, "because it calms my mind" and helps him sleep. Judging from the message boards and the people I've talked to, this flexible you-do-you approach to microdosing is where many people end up.

Microdose Consultants

Not everyone is going to experiment with their neural pathways, not without some kind of expert to oversee their progress. For those who need advice, feedback, or personal attention in their microdosing practice and can afford it, there is a thriving community of consultants a click away.

As interest in microdosing has expanded, so have a host of related services, from microdosing books to online courses to coaches. It's not hard to find microdose gurus, alchemists, consultants, and "pros" of various sorts, with a variety of credentials and skill sets. Whether or not their service justifies the often high fees associated with consulting is another matter. When it comes to microdosing, unless the consultant is supplying the mushrooms, you may just be paying for hand-holding.

But a lot of people seem to want their hands held, because microdose consultants are prolific. They run the gamut from well-known online coaches like Paul Austin, who started his business in 2017, charging $97 a session, but five years later, his microdosing program costs $1,000–$2,000 a month, with a minimum three-month commitment, to smaller operations like Zenchronicity, based in Denver, where two sisters suss out which chakra contains your trauma and then prescribe certain *P. cubensis* cultivars they believe address said chakra.

There are more choices in the Netherlands, where magic truffles are legal. Like House of Oneness, which offers a kind of plant medicine everything salad, including a thirty-one-day microdosing video challenge for €555 (about US$600 as of this writing). Microdose Pro, also in the Netherlands, provides both private and group sessions, in person or online, and the aforementioned Microdosing Institute has an online guided Zoom program for $1,000 (but doesn't offer the drug). Microdose coaches may be found through directories like the Third Wave website, which lists coaches from the US, UK, and Costa Rica, or by word of mouth. I once met a neurologist who offers microdose consulting, but she doesn't advertise. The only way you would find out she performs these services is if it came up in conversation.

Navigating the surfeit of certificates that serve as credentials for coaching is, like guiding, mind-boggling. A microdosing coach may seem expert, and there are certainly people who have lots of experience, but there is no generally recognized certifying body. Instead, many coaches boast certificates in yoga, meditation, breathwork (in particular, the Wim Hof Method), and massage. Moving into the more esoteric, you might encounter certificates in the Energy Blueprint (an educational website, online course, and supplement store dedicated to reversing the effects of chronic fatigue syndrome) and ANS Rewire, a "brain rewiring" course, or connections to Mazatec healers. Organizations like Third Wave offer microdosing coach training as part of their general coaching certification program. According to the Third Wave website, over three thousand people have enrolled in its course—four enrolled while I was scrolling through—and completion ensures you will receive good placement on their directory of coaches. At the Microdosing Institute, you must first complete the six-week microdosing program, then you can go on to learn to be a coach, for €4,797 (about $5,250). Microdosing, or being a microdose coach anyway, can be mega-expensive.

And there are also apps, which feel a lot like diet apps. You do all the microdosing, and the app helps you keep track of your behavior. The Psily microdose optimization app helps you chart your mood, energy, and sociability, as well as "tracked stacks" like the mixes of supplements and psilocybin you take; Microdose .me is a tracker affiliated with Paul Stamets (he announced the app on *The Joe Rogan Experience*) that has over thirty-three thousand subscribers, a potential gold mine of anecdotal data. And indeed, an analysis of their microdosing habits and experience was published in *Scientific Reports*. If you prefer to read and write with a pen and not your thumbs, there are books, too. For example, Jaden Rae's *Microdosing Guide and Journal* combines dosing techniques and a format for charting your day-by-day physical and mental reactions. There are, in fact, quite a few microdosing journals on the market. Likewise, there are multiple psilocybin microdosing guidebooks, but most that have recently hit the market seem to be slender self-published works that, based on their reviews, end up pissing off buyers.

When it comes to microdosing, most information is anecdotal. A coach may not have that much more sage advice than a person named Educational Tomato on Reddit. Without legal access to the drug and doctors who can safely prescribe it, we are in a professional vacuum.

I have my own small contributions to offer to the body of anecdotal knowledge. On my optimal protocol—150 milligrams of Golden Teacher every third day—I had no side effects: no nausea or dizziness, no crabbiness or negativity. At first, my short-term memory was not improved; in fact, I might have been a bit flaky and distracted. I had to search for synonyms and anecdotes that lingered on the edge of my recollection: it was like generally remembering where you parked the car but still having to walk around to locate it. I questioned if I was better or worse off. But the next day, my first day off (and a day that many users describe as the best day of their protocol), I dropped right into the flow state and felt very positive about, well, everything. Even a battle over the car keys with my ninety-six-year-old, sight-impaired father was less nerve-jangling than such encounters had been in the past. The day after that, when the drug and afterglow had cleared from my system, I felt similarly uplifted—on target professionally and generally more empathetic. Instead of instantly feeling annoyed over the condition of our coffee station after my husband left for work, I experienced a moment of objectivity that allowed me to realize I didn't really need to care about the mess. And that state of mind continued for as long as I continued to microdose. In fact, as I neared the end of my trial, my husband said he hoped I'd keep microdosing, I guess because I was less pissy. I just felt focused and mellow. It was wonderful.

I found one consistent side effect, regardless of dose size, was warped time perception. For example, sitting at my computer for six hours felt like three. That's not surprising to the time perception researcher Marc Wittmann. "Now you could say that a full-blown trip will for sure affect temporal processing, as it affects so many cognitive, emotional, etc. domains," he wrote me in an email. "But . . . most extraordinary, sub-threshold doses of psilocybin similarly affected time perception."

I had another surprising effect. One of my problems with long stints at the computer is neck pain. I bought a standing desk, but I still sit at it, so I bought an ergonomic chair. I am constantly adjusting both pieces of furniture, as well as my monitor angle. But on the microdosing protocol, I had no neck pain, even though I was working as long as or longer than in the past. Now I am thinking the problem was never my apparatus. It was me. Often when writing, I am very tense, my shoulders up near my ears, my chin jutting into the light of my computer screen. I have a low-grade buzz of anxiety: *Am I missing something? Am I getting this right? Do I sound like an asshole?* It's like I am in a constant fight-or-flight response. But somehow, that didn't happen on the microdoses. One of the ways I think the mushrooms enhanced my work experience was not by what they added, but by what they subtracted: the anxiety that was inhibiting me.

9

Therapeutic Trips

Liana Gillooly, the strategic initiatives officer at the Multidisciplinary Association for Psychedelic Studies (MAPS), her gown sparkling in the dusty sun, told her Burning Man festival audience, "Forget microdosing. More relief can be found by doing one or two large doses." She's referring to the kind of relief you've been reading about in the headlines: relief from neuropsychiatric disorders like addiction, depression, PTSD, and anxiety.

One or two large doses (about 20–30 milligrams of synthetic psilocybin) is pretty much the standard that patients receive in clinical trials. Those who choose to self-medicate at retreats and in private sessions might take an equivalent average high dose of around 3 grams of dried mushroom. Headlines like "Psychedelics May Ease Cancer Patients' Depression, Anxiety" and "Magic Mushroom Hallucinogen May Treat Problem Drinking" are generated by studies of patients who have taken a dose this size, hoping to relieve symptoms of their disorder. These are therapeutic trips, and they are the primary interest of scientific study, much personal experimentation, and pharmaceutical ambitions.

Psychedelic Pharmacology

Today, the landscape of Psychedelics Inc. is populated by hundreds of private companies. That may sound like a lot, but in terms of value, the entire sector doesn't even equal 2 percent of the pharmaceutical giant Pfizer. There was a burst of investor optimism during the COVID pandemic, but that has slowed down, and companies have started to cut programs, to merge, to align, to divvy up the territory. The biotech company Compass Pathways is dominant in psilocybin research.

Whether you are trying to develop a new statin or a new, patentable version of the psilocybin molecule, drug discovery and development are the costliest part of the venture. Compass is developing patented molecules like COMP360 psilocybin, a "synthesized psilocybin formulation," basically, psilocybin that has been crystallized in water and mixed with a tableting agent. They are financing studies researching how COMP360 might affect patients with treatment-resistant depression, and accordingly, whatever results emerge from the studies will be tied to their patented psilocybin product. (Purely academic studies will use the generic synthetic psilocybin first isolated by Albert Hofmann at Sandoz.) Other companies are pursuing psychedelic drug development that includes microdoses, shorter-acting trips, and drugs that skip the trip part altogether by evaluating which aspect of psilocybin's chemical structure is associated with mental health benefits and removing the hallucinatory piece.

But psilocybin is a very different animal from other drugs. So far, it has shown to be most efficacious if coupled with therapy. There is no precedent in our health system for a prescription drug tied to psychotherapy. The problem isn't producing the drugs to scale; it's producing therapeutic services to scale. Regulators would need to determine how much therapy is the minimum allowable, a tricky endeavor, said Ben Sessa when we talked, because the nature of psychotherapy is driven by individual presentation and a "collaborative discussion with the patient." A lot more therapists and social workers would need to be trained in psychedelic treatment. (Currently, there are almost two hundred thousand practicing therapists in the US, but more than twenty-one million Americans suffering from depression.) As Mark Plotkin joked, kind of, "If my kids asked me what business to get into in the future, I'd tell them tattoo removal or psychedelic therapy."

Issues of insurance coverage, equitable availability, screening and diagnosis, convenience, safety, and transitioning patients off medications or from one to another are the kinds of problems that need to be addressed if psychedelic therapy is to go mainstream. But these are ultimately good problems to have if the research can substantiate claims that psilocybin heals.

And to that end, researchers are casting a wide net into neuropsychiatric disorders with both synthetic psilocybin and psilocybin-like products. They are looking at depression, anxiety, and addictions, but also headache, chronic pain, and eating disorders. And more questions are being asked. Will psilocybin help gamblers? People with long Lyme? Alzheimer's? And if, as the preliminary research suggests, psilocybin does indeed address a wide range of disorders, then maybe the reason is that the drug is addressing a common denominator among them all.

Mushrooms and Addiction

Daniel Kreitman, an affable furniture upholsterer from Maryland, sometimes dreams of smoking. For thirty years, he smoked one to two packs of cigarettes a day, but that was ten years ago, he told me, before he participated in a Johns Hopkins clinical trial for smoking cessation. The trial consisted of two months of preparation—talk therapy, journaling about his smoking habits, and guided meditation with Mary Cosimano, one of the most experienced trip guides in the country—then three psilocybin trips a couple of weeks apart, followed by debriefing with the doctors, Albert Garcia-Romeu and Roland Griffiths. He smoked throughout the preparation phase.

In anticipation of his trip, Kreitman was asked to make up a mantra—essentially, an intention—"something I could say to myself. It was, 'For my family and myself, I will quit forever.'" In a small room at Johns Hopkins, the researchers had him crush his cigarettes in a wooden chalice and repeat his mantra. After taking a capsule of psilocybin, he lay down on a couch with an eye mask and headphones playing a musical soundtrack "that kind of took you along your trip. Let you go different places, like a winding road." His sessions were eight hours, and there were two people in the room with him, Dr. Garcia-Romeu and Cosimano, who periodically took his blood pressure and asked him how he was doing.

For his last trip, Kreitman was allowed to include some of his own music, including Aaron Copland's *Appalachian Spring*, and during the music, "I started leaving the flowers and colors and went into the skies and stars and infinity. I left this whole thing, whatever this is." He has never smoked since. I asked him if it was because the experience was so beautiful that he just felt grossed out by tobacco. Nope, he said, it was like he had never been a smoker in the first place. Not only is he tobacco-free ten years later, but "whenever I hear *Appalachian Spring*, I get kind of emotional. Literally, right now, I am tearing up a little bit. It's that deep."

Kreitman was part of a pilot study that had fifteen participants. Six months after the study, 80 percent remained smoke-free. That number declined to 60 percent after three years. A pilot study like this is more an indicator of whether further study is warranted, and it was. More current research, by Peter Hendricks of the University of Alabama and others, has found psilocybin may be more effective than the nicotine patch. He told me, "If I give one hundred people the best treatment and thirty quit, that would be standard. My hope is with psilocybin, maybe we can bump that up to fifty or sixty who quit. It would still be a boon for public health."

Psilocybin may be effective at treating other addiction disorders, too. Since the mid-twentieth century, researchers have studied how effective psychedelics are at curbing alcohol abuse. That work continues today. In a 2022 randomized double-blind study of people with alcohol addiction, psilocybin—in conjunction with psychotherapy—produced "robust decreases" in heavy drinking days. Anecdotally, trippers have long used mushrooms to get sober, though their psychedelic use can run counter to twelve-step programs. I spoke with a middle-aged farmer who has suffered from depression his whole life. An alcoholic many years sober, he uses psilocybin to mitigate the symptoms of depression that led to his drinking in the first place. But he was distraught because he couldn't tell his AA sponsor that he was using mushrooms, which constitutes a lapse. It didn't matter that Bill Wilson, AA's founder, used psychedelics.

But alcoholics who use psychedelics aren't necessarily out in the cold. The organization Psychedelics in Recovery (PIR) is a fellowship of people in twelve-step programs who also have an interest in psychedelics as an aid to that recovery. PIR offers scheduled Zoom meetings that include people from all over the world, subdivided into groups like Women Beyond Sobriety, Queer in Recovery, and Men's Meeting, and in-person meetings in a variety of cities across the US.

There are other addictions under study. Dr. Hendricks has been looking at psilocybin-facilitated treatment for cocaine use. He explained to me that while the study is incomplete, the data so far indicates that over time, more users abstain, and lapses don't automatically lead to relapse. I asked why he thought this was, and he shared a most remarkable analysis. "Imagine you are on a diet," said Dr. Hendricks, "and you've been doing well, but you have a piece of pie at a party. And you think, *I've blown it. There's something wrong with me. I can't change.* At the heart of that is a sense of shame. A better response is guilt. *I made a mistake, I feel bad about it, and I want to learn from it.* I think there is something about the psychedelic experience that is likelier to react with guilt, maybe convert shame into guilt. Guilt keeps you from repeating your mistake; it's focused on behavior. Shame is feeling bad about who you are as a person, in which case, what's the motivation to change your behavior?"

Gambling disorder, compulsive sexual behavior, binge eating, and internet gaming disorder (which it seems half my son's high school class suffered from at one point or another) all may get the psilocybin treatment. If it works for tobacco and opioids, cocaine and alcohol, why not gambling? It's easy to disregard these types of disorders, idling as they do on the fringe of more prevalent addictions, but they have real-life consequences. I had a friend whose husband's addiction to mah-jongg led to their financial ruin and divorce. It's ironic that psychedelics—a Schedule I drug—seem to help change behaviors society has traditionally condemned as moral failings.

Mushrooms and Anxiety

Small studies have suggested psychedelics are particularly effective when used for anxiety-related disorders. Post-traumatic stress disorder (PTSD) is an anxiety disorder that occurs following traumatic experiences. Anorexia nervosa (AN) is another kind of anxiety disorder, as is obsessive-compulsive disorder (OCD). I met a young man who had suffered from trichotillomania, or hairpulling, for years. He would feel a buildup of anxiety that became increasingly stressful until he plucked a hair. His anxiety would be temporarily assuaged, like letting a bit of air out of an overfilled balloon. But then the anxiety would return, it would fill him back up, and he'd have to pull out another hair. Over time, he denuded his scalp and eyebrows. He told me it was embarrassing and isolating. While trich can be managed with tracking one's pulling behaviors and identifying triggers, there is no known cure. But this individual's condition resolved after taking a self-administered psilocybin trip. He was reluctant to share the details of the trip, the dose, or the frequency, but when I met him, he had been a few years free of hair pulling. He said the compulsion stopped when his anxiety calmed after consuming the mushrooms. Sometimes he dreams he is pulling again. When he wakes up, he is so very relieved to realize he is not.

OCD is another disorder that may respond to psychedelic therapy. People with OCD often hide their symptoms, and as a result, the condition is probably underdiagnosed and undertreated. Obsessive-compulsive disorder is a kind of wicked feedback loop where intrusive and obtrusive thoughts or urges, often associated with anxiety or discomfort, are acted upon to neutralize the distress, which just reinforces the compulsion. A classic example is an obsession with contamination by microbes or germs, causing the sufferer to wash their hands constantly. Washing their hands reduces the anxiety, but it also reinforces the compulsion that washing their hands protects them from germs. It can be debilitating, according to Christopher Pittenger, cofounder of the Yale Program for Psychedelic Science, whose research includes studies on OCD and psilocybin. The classic treatment is to try to interrupt the loop with antidepressants or cognitive behavior therapy—which works 30–40 percent of the time, said Dr. Pittenger at the Horizons psychedelic conference in New York City in 2022. Preliminary research has shown psilocybin may be more effective, and those patients who felt their symptoms relieve within forty-eight hours stayed better for three months. While the patients say they still have the obsessions, they just don't matter anymore. "The difference isn't the nature of the thoughts," said Dr. Pittenger, "but one's relationship to the thoughts—whether they are seen as powerful and important." OCD sufferers I have spoken with describe a similar qualitative change: the intrusive thoughts lose their power and are therefore easier to manage.

At Blue Portal, the retreat I visited in Jamaica, I made friends with Meg Maginn, a therapist who specializes in patients with eating disorders. She had a critical patient, a nineteen-year-old girl whose severe anorexia had put her into the hospital multiple times. Meg was eager to find anything that might help this wasting girl, and she came to the retreat to try the mushrooms herself. "People with eating disorders associate foods with phobias," she told me. "They *know* they won't get fat if they eat that Oreo, but they *believe* otherwise. I think psilocybin breaks deep-embedded belief systems, which disrupts rigid patterns of thought." Anorexia nervosa is a life-threatening psychiatric condition with a high relapse rate. It's often accompanied by anxiety and depression, PTSD, or autism spectrum disorder. Trauma is a risk factor, especially sexually related trauma, which may explain why the disorder is more common in females. Neither antidepressants nor antipsychotics have much impact. The most widely used treatment is cognitive behavioral therapy, but dropout rates are high. Psilocybin may help these patients deal with their depressive symptoms and address their eating disorder by softening their resistance to recovery. The science is at a very early stage, but promising. Both Johns Hopkins and Imperial College London are conducting pilot studies on psilocybin and AN. "The question is not so much how to treat anorexia," said Meg Spriggs of Imperial College London, "but how do we help people engage in the process of recovery." Outside the clinical trials, AN sufferers have treated themselves with mushrooms. Their testimonies explain how they were able to see themselves from an objective point of view, which lent an appreciation for their physical body that they had not felt before. Especially for those who cannot afford therapy, mushroom self-treatment can be a boon. And ongoing online support, however erratic, is available within the community. As one fellow wrote in response to an AN sufferer's trip report on r/shrooms, "Just stay aware in the following months. . . . You've done the hard part . . . now it's a case of staying on top of it."

PTSD and Mushrooms

Post-traumatic stress disorder (PTSD) is the most studied anxiety disorder that may be treatable with psychedelics. Generally triggered by a terrifying event, PTSD is characterized by flashbacks, nightmares, and severe anxiety. Those symptoms can happen to anyone after a traumatic event; in the case of PTSD, they persist. PTSD is like an injury to one's ability to process emotions. It is usually treated with psychotherapy, antidepressants, or antipsychotics, but success is limited. Some patients turn to substance abuse; others develop anger management issues or even die from suicide.

There are many reasons why someone might suffer from PTSD, but it is the community of veterans that is spearheading the demand for psychedelic access. Seven percent of all vets suffer from PTSD. Maybe that's the result of the "endless wars" in Afghanistan and Iraq; more vets have died from suicide since 9/11 than have died on the battlefield. But the federal government, according to Lieutenant General Martin Steele, an advocate for psychedelics to treat PTSD, has been slow to fund research on the mental health of soldiers. Indeed, the government didn't even acknowledge the epidemic of PTSD until recently.*

* It's not like the condition appeared suddenly: during World War I, what we call PTSD today was known as *shell shock*. Those soldiers, like today's, exhibited "flashbacks and panic attacks evoked by situations reminiscent of their traumatic experiences" in combat.

Research into psychedelics and PTSD, which was dominated by LSD before prohibition and MDMA since, suggests psychedelic-assisted therapy can improve symptoms, even lead to remission. The efficacy of psilocybin in treating PTSD is being studied as well. But MDMA has led the field in research on vets, mostly due to the work of the Multidisciplinary Association for Psychedelic Studies. In 2020, the FDA expanded access to MDMA for clinical study, and full approval is expected in 2024. A complementary study in Europe may similarly be approved by the European Medicines Agency in 2026. Both MDMA and psilocybin are thought to decrease the activity of neural pathways in the brains of PTSD sufferers that trigger fear and anxiety, leading to a reduction in their fear and anxiety symptoms.

Veterans don't have a monopoly on PTSD. Folks in other high-stress, public-facing jobs, like police officers, clergy, firefighters, social workers, and ER staff, might benefit from psilocybin therapy, too. Throughout the shutdown stage of the pandemic, here in New York City, we banged our pots and pans out the window every evening at 7:00 p.m. to show our support for first responders, but we couldn't imagine what they saw and heard. A freshly minted MD I know sent this report of a day in a New York–area hospital in April 2020. "I walked in on a code [shorthand for cardiopulmonary arrest], and then halfway through rounds another code. It seemed like there was a code announced on the loudspeaker every 30 minutes after that. COVID ICU patients out in the open everywhere on stretchers, all intubated and doing terribly." As one resident said in a moment of respite, "I have no admissions and I hate it because then I'm left alone with my thoughts." Another MD reported, "We coded three people in their 30s this past week. One was an anesthesia resident (he's the only one who survived). I felt so sick during that code because it hit so close to home. My attending later told me, 'I almost threw up during that. I will 100% have PTSD.'" A current study at the University of Washington is looking at the ability of psilocybin to help these clinicians with symptoms of depression and burnout related to their frontline COVID work.

Mushrooms and Chronic Pain

"I have chronic back pain," wrote one person on Reddit. "For the duration of my trips on LSD, MDMA, and mushrooms, I am pain-free." Anecdotally, trippers have reported that while tripping, they are relieved of back and neck pain, fibromyalgia pain, migraine, even hemorrhoid pain. The trip gives them a "vacation from pain." This has been my experience, too. On the other hand, psychedelics can amplify some people's pain, so the mushroom can't be seen as a panacea. But when it works, here's how psilocybin may be helping.

The transition of acute pain to chronic pain involves changes in nervous system structure and function, which could affect the way your brain and spinal cord process painful (and non-painful) signals. What may be going on is that constant pain signals due to injury or illness could be strengthening certain neural pathways that result in the physical and emotional perception of chronic pain after the instigating injury has healed. Using psilocybin may reset those problematic neuropathways.

Several promising surveys and studies were published prior to the current psychedelic era on psychedelics and chronic pain, primarily on phantom limb pain, cluster headaches, and cancer-related pain, and more have been conducted since 2006. All of them showed the promise of a psychedelic solution. Fadel Zeidan is working on a study of phantom limb pain, which occurs, to varying degrees of intensity, in around eight out of ten people who lose a limb, even if the amputation was planned. Although the data is not available yet, Dr. Zeidan thinks psilocybin might work to reduce chronic pain because it may impact areas of the brain involved in self-reference and pain. "In individuals with chronic pain, the moment-to-moment experience is contaminated by pain. The pain is not necessarily separable from who we identify as self. A chronic pain patient identifies *as* pain. We think [psilocybin] may be able to decouple the pain experience from the self."

Another type of chronic pain is migraines and cluster headaches. These are distinct headache disorders, each with characteristic debilitating attacks of head pain accompanied by other neurological symptoms. An exploratory controlled study in 2021 by Yale's Department of Neurology and the Veterans Affairs Headache Centers of Excellence found a single dose of psilocybin had an enduring effect on migraine headache. I asked the paper's lead author, Emmanuelle Schindler, why she thinks it worked. "I'll be spending the rest of my career working on that," she said, but very generally, she thinks psilocybin may target the underlying pathology of headache disorders, allowing for lasting effects after a single dose.

Cluster headaches, on the other hand, are short, intensely painful, and reoccurring in brutal cyclical patterns or cluster periods, usually at night. As one person on Shroomery described it, "The beast comes to play every 2 years in the height of summer and dances with me for 3 to 4 weeks." People who suffer from them—about one in one thousand in the US—call them *suicide headaches*: they are so bad you want to die, and one man actually did shoot himself in the eye to stop the headache. (He survived, incredibly, without neurological deficits, but his cluster attacks continued.) Psilocybin's ability to reduce the frequency of cluster headaches first came on the scientific radar because of the chat group Clusterbusters, where cluster headache sufferers shared their DIY remedies. That led to surveys, which anecdotally showed psilocybin was indeed effective (as is LSD). Cluster headache sufferers have reported that on psilocybin, they feel a distinct action, a popping or opening feeling. One man described this sensation as occurring on one side of the face, where, maybe coincidentally, lies the upper branch of the trigeminal nerve and which is associated with sending pain, touch, and temperature sensations to your brain and face. But like microdosing, the efficacy of psilocybin use for cluster headaches remains anecdotal.

Chronic pain is also an aspect of post-treatment Lyme disease syndrome (PTLDS), or long Lyme, which occurs in 5–15 percent of patients. The original infection is caused by a bacterium, *Borrelia burgdorferi*. Usually, the disease, of which there are twenty thousand to thirty thousand cases a year, is treated successfully with antibiotics. But some patients have ongoing symptoms, including fatigue, brain fog, depression, and musculoskeletal pain, what one woman described to me as "post-infectious fuckery." A pilot study currently underway at Johns Hopkins is looking to see if psilocybin can affect PTLDS symptoms, with the goal of improving patients' quality of life. It's a proof-of-concept study, the concept being that maybe long Lyme is a neural construct like other chronic pain conditions.

But in the meantime, there are people who will pursue treatment regardless of the state of the science, as in the luxurious Eleusinia Psilocybin Retreat in Mexico, which uses the promise of psilocybin for chronic Lyme in their sales pitch.*

If chronic pain—and not just those conditions I've mentioned but other similar conditions, like long COVID—can be attributed to the sufferer's perception of pain, and if psilocybin does indeed change that perception, then maybe relief is at hand. What's more, in the forums, numerous COVID patients have reported that after a psilocybin session (both macro- and microdoses), they regained their sense of smell and taste.

Mushrooms and Depression

Depression is one of the most common mental disorders and one of the most difficult to treat. Costly, too. My mother went to an analyst for treatment five days a week for twenty years. She told me how once, when she was entering his office building and encountered him on the sidewalk, he pointed to an expensive car parked at the curb and said, "You bought this."

An estimated twenty-one million adults in the US had at least one major depressive episode in 2020—over 8 percent of the population. The promise of psychedelics to help mitigate symptoms of depression is maybe the most compelling aspect of psychedelic research, from both an economic and a compassionate standpoint. And it *is* promising. Multiple, mostly small,

* And sometimes a cure can be had unintentionally. According to *Business Insider*, a woman mistook LSD for cocaine and ended up taking 550 times the normal dose. She tripped for thirty-four hours. After that, she microdosed. In her estimation, it cured her chronic Lyme. It hardly needs to be said this is not a good idea to do at home.

studies have shown increases in antidepressant effects, in some cases enduring effects from one to two moderate doses of psilocybin, and as early as 1950, psychologists and psychiatrists noted that psychedelics could be used to shorten psychotherapy.* In 2018, psilocybin received its first breakthrough therapy designation from the FDA as a potential treatment for depression. This is important because it simplifies procurement hurdles, destigmatizes the drug, and basically expedites its path to becoming an FDA-approved medication. It may be as effective as the SSRI Lexapro; the big difference is you don't have to take psilocybin every day.

But when the psychedelic company Compass Pathways conducted a larger and more rigorous study and published their results in late 2022, the findings were a reality check. The Compass study looked at people with treatment-resistant depression—that is, people who have tried all the current available medications to no avail. Three groups each took a single, but different size, dose of the company's psilocybin product, COMP360. Depression symptoms declined in all three groups, with the best results in the highest-dose group, but at the three-week mark, only 37 percent of those high-dose participants saw continued improvement, and over the following weeks, that improvement decayed. After three months, only 20 percent of the high-dose

* So how do you measure depression? There are no biomarkers (although mental health issues have been associated with the gene mutation MTHFR; but having the mutation doesn't mean you will suffer from depression—it mainly causes a failure to convert folate into the more bioaccessible L-methylfolate, which plays a role in making neurotransmitters like serotonin). It's hard to measure a feeling, especially since feelings can be episodic, and then determine if that feeling measurement is accurate. However, there are a few assessment scales. One, called GRID, scores symptoms. Another is QIDS-SR-16, which stands for Quick Inventory of Depressive Symptomatology. These are self-assessing measures of depression severity based on symptoms like sad mood, self-criticism, guilt, and suicidal ideation. These measures don't capture the complexity of depression, but they are the best we have right now.

participants still felt substantially better. Nevertheless, even though the number of people who benefited was smaller and the effects of shorter duration than hoped for, the study showed that a single dose of psilocybin still offered relief for some of these very-difficult-to-treat patients.

Depression comes in many shapes and sizes, and in addition to research on psilocybin and treatment-resistant depression, studies are underway to look at depression in combination with bipolar disorder, suicidal ideation, and anxiety; depression arising from cancer, Parkinson's disease, alcoholism, Alzheimer's, and melanoma diagnoses; depression brought on by palliative and hospice care; and demoralization from long-term AIDS. But of course, many people don't feel like they can wait for research findings and are treating themselves, often doing their integration work with their own therapists or in small groups of like-minded people.

The trip reports of folks who are self-medicating show both the incredible promise of psilocybin to relieve symptoms and also a wake-up call regarding its limitations. The general anecdotal consensus I've observed is that large doses (in the 3-dried-gram range) don't cure depression—though it seems to help many people feel relief from their symptoms for weeks and months at a time, after which they trip again. But mostly, trippers report that psilocybin helps you work on yourself to treat your depression. "[Psilocybin] may motivate you to do things you've been putting off, give your perception a boost to increase your confidence, or give you the gratitude required to stave off nihilism," wrote one person who goes by the name Socrateshroom. "But without the work (integrations, personal reflection, and action) they are like any other tool used incorrectly."

Mushrooms and Grief

Most of the patients who have attended the Jamaica Grief Retreats are moms who have suddenly lost a child to accident or suicide and are trapped in unremitting sadness. Grief isn't mental illness, but complicated grief, when it carries on for years and affects one's ability to function, is a diagnosis that may be treated with cognitive behavioral therapy or drugs. It may also be treatable with psilocybin and the companionship of other grieving parents.

Dingle Spence, a retired palliative care physician in Jamaica, has conducted group retreats for these parents. She explained to me that the group format works because "everyone there completely gets what you are dealing with." The patients "double bond" during the retreat, first as parents burdened by similar tragedies and second as fellow travelers on the psychedelic journey. Many, said Spence, stay in touch long after the retreat is over. Jamaica Grief Retreats administers "full journey doses" of 3–7 grams of dried mushrooms, depending, of course, on the person. The highly trained staff includes palliative care doctors, death doulas, and psychotherapists with bereavement experience. The training is key, suggested the psychiatrist Neil Hanon at an End Well panel discussion (End Well is a nonprofit dedicated to improving end-of-life experiences), because mushrooms may retraumatize grieving parents if they are taken without wraparound, grief-oriented therapy. "It's not like you take mushrooms and your grief is gone," he said.

I read many trip reports about psilocybin and grief, and one persistent theme was the relief—however bittersweet—that comes from realizing that there is no such thing as closure in this life. One of the parents who participated in the End Well panel discussion said, "When you are in that kind of pain, you want to move away from the pain," and doctors are quick to prescribe drugs like Prozac. But, she said, psilocybin helped her face her grief. "It's probably the hardest thing I've ever done, but in the long run, it helped me live with the pain." Jamaica Grief Retreats had only conducted three retreats at the time I spoke with Spence, and they are considering how to address a variety of grief needs, like those of the siblings and spouses left behind, though, sadly, there is no shortage of bereaved parents.

Science would like to understand how and if psychedelic therapy relieves complicated grief, which shares symptoms with PTSD. It's possible our understanding of PTSD can inform what is going on with people suffering from complicated grief. In an American Brain Foundation webinar, "Healing Your Brain After Loss," the neurologist Lisa M. Shulman, author of *Before and After Loss: A Neurologist's Perspective on Loss, Grief, and Our Brain*, said PTSD is understood as resulting from any emotionally traumatic event, like the death of a child.

From the brain's perspective, traumatic loss is perceived as a threat to survival, and it reverts to survival mechanisms, which subsequently change body functions like memory, sleep, even physiology, as in the case of Takotsubo cardiomyopathy—so-called broken-heart syndrome. "After loss, our brain actually rewrites itself because stress is a very potent activator of neuroplasticity," she said. "Neurons that fire together, wire together." When repeated firing is reinforced, it becomes a default setting. And that default setting, that maladapted neuroplasticity, can manifest as a state of fear and anxiety, or PTSD. Since psychedelics are increasingly shown as effective in the treatment of PTSD, Dr. Shulman thinks they might similarly help those suffering from the pain that defines grief.

Consider the story of Virginia, who lost her adult daughter to suicide. The young woman, who was outwardly successful and independent, had effectively hidden her mental state and her intentions from family and friends throughout the COVID lockdown. Like so many other parents in her position, Virginia became subsumed with guilt, racking her brain for signs she might have missed and remembering comments like "I'm never having kids," which in retrospect took on a different light. Virginia had no experience with psychedelics, but "people came out of the woodwork" to recommend she try them, and so she found a family member who agreed to procure the mushrooms and keep her safe. She did not engage in professional integration therapy. She did the work herself.

In the session, Virginia did not see her daughter, but she encountered an array of people she knew, people who were experiencing all sorts of problems that they expected her to resolve. "And it came to me that I couldn't help them. It was beyond my control to help them. And somehow that was tied to my daughter." Ultimately, she accepted that she couldn't have stopped her daughter from doing what she did, and that realization helped her let go of the guilt. "I went from a place where I couldn't continue to a state of just grief. But I have more work to do."

She is planning another trip.

Mushrooms and Palliative Care

When age or disease takes us, that journey is one we traditionally do not take alone. Many groups, both indigenous and nonindigenous, practice rites of passage, ministrations upon their deathbeds, and proscribed funerary protocols. They know what's going to happen to their body and, depending on the state of their soul, what will happen after they are dead. What the variety of religious and spiritual rites around death have in common is they help the dying change their narrative from a personal story—*This is happening to me*—to a communal story—*This happens to us all.*

For so many of us, however, death is a medical event, experienced isolated in our wards, awaiting fateful news. When faced with the horizon of one's own demise, the temporariness of this container can lead to feelings of pointlessness and depression. "Generally speaking," explained Kevin Roux in a conversation—he's a hospice-registered nurse case manager—"when patients are given a life-limiting prognosis, they spend a significant amount of time trying to make sense of their lives. What does their life mean? What are they going to miss?" By and large, he noted, psychedelics can lead to people experiencing their own consciousness so differently that the fear of death dissolves. "They say, 'I now know I am not my body. My consciousness feels separate from my body.'" Liberated from fear of death, patients may even be curious about death and what lies beyond.

Since the mid-twentieth century—excluding the scientifically fallow war-on-drugs era—psychedelics have been studied for their ability to mitigate cancer-related demoralization, specifically feelings of loss: loss of meaning, loss of hope, and loss of purpose. It's some of the most persuasive science in the field, and more current study is suggesting the same. In 2016, studies from NYU and Johns Hopkins found that most participants with terminal cancer showed substantial and sustained reductions in depressive symptoms after treatment, and up to six months afterward. Those studies have been reinforced by even more recent research that shows the mushroom relieves existential distress by means of "mystical-type experiences." While participants in the studies variously describe feelings of gratitude, forgiveness, death transcendence, and increased prosocial attitudes, in both studies, the researchers found it was the mystical experience that helped these patients feel better. As the religious scholar Huston Smith wrote, their pain might continue, but not in a way that matters, "so completely is it set in cosmic perspective."

But there is no guarantee what kind of trip anyone is going to have, just as there is no guarantee how a surgical procedure is going to turn out. "We must be careful about the language we use when preparing people for this experience," said Roux, who advised the state of Oregon on psychedelics and end-of-life patients. "We prepare people for surgery all the time by informing them of the possible risks and benefits of the procedure. We can do the same with psychedelics; if an experienced guide can give you a clear picture of the risks and harms associated with psychedelics, you can mitigate those risks. When everything is done right, from a well-prepared set and setting to the right dose, it is reliable to expect a low incidence of harm."

What would it look like to provide psychedelic care to end-of-life patients and potentially their families, too? One model might be psychedelic services offered within oncology centers. The Aquilino Cancer Center in Maryland is a kind of all-in-one facility for cancer patients that encompasses laboratories, chemotherapy, radiation, surgery, rehabilitation, and integrative medicine (like guidance on nutrition and diet). It also is home to Sunstone Therapies, a company that is studying psychedelic therapies in a medical setting. I spoke with Sunstone's Brian

Richards, an engaging clinical psychologist and researcher, and the son of psychedelic medicine pioneer William Richards. "We've got cutting-edge medicine," he told me, "but the wrenching existential suffering people go through, quietly and privately, is rarely addressed." Psilocybin could help patients in palliative care, and their families, deal with the emotional issues that block them from living with greater acceptance of their diagnoses.

He and his colleagues are working on a prototype, the Bill Richards Center for Healing, where they hope to help people who face traumas of all sorts, from losing one's breasts, to the fear of not being present for a kid or spouse, to a fatal prognosis. "If we thoughtfully develop the therapy envelope," said Dr. Richards, "then people may have experiences that can be transformative." He told me the story of a musician with "stable" stage IV lung cancer who had a tough trip but in integration therapy found his own kind of peace. "He said, 'I died again and again and again until death itself was boring.'"*

Hospice Care and End of Life

When treatment options become ineffective or too burdensome, or if a patient gets to the point where she says, "I'm done," she may be eligible for hospice services. From that point,

* There are many hurdles to using psilocybin within our already established medical model, but dropping psychedelic services into the "medical envelope" of cancer care centers might work because they already employ staff with multiple disciplines, including nurses who are especially qualified for the job of sticking with a patient for many hours. A cancer center will already have experience with complex medicines—and, one would hope, how those medicines might interact with psilocybin or some other psychedelic—and the administrative infrastructure to deal with the paperwork and accountability.

the main objective is keeping the patient as comfortable as possible. In the hospice scenario, patients forgo aggressive life-prolonging interventions and focus on quality of life, and psychedelics might help them make peace with the inevitability of death. Stephanie Barss is a family and psychiatric mental health nurse practitioner in palliative and hospice care in Oregon. She is participating in a psychedelic harm reduction and integration program in anticipation of the day when psychedelics can be legally used to help folks with advanced or life-limiting illness. "We medicate death," said Barss. "One of the primary reasons I went into this is because once in a while there is someone who suffers, and though we give them narcotics, they are still in emotional agony. It's not bone cancer or kidney failure. It's existential distress. When I give that kind of person morphine, I feel like I am doing a disservice because some people may have unresolved stuff that they could possibly resolve—some of it anyway—and as a result die with less suffering for them and their families. Ultimately, it's about finding peace."

The end-of-life studies generally employ 25–30 milligrams of synthetic psilocybin, though underground use may employ more. Aldous Huxley famously had his wife administer a high dose of LSD hours before his death. But dosing someone in their final months of life is tricky because there are so many variables, including their catabolic state—at what rate their body is breaking down—that would make it difficult for the practitioner to determine a dose "that can allow for a transcendent moment," said Roux. Indeed, a patient who has abused alcohol for twenty years might require a much, much larger dose due to chronic brain inflammation. But beyond the physical inhibitions, it might be emotionally challenging for terminal patients to experience ego dissolution. "It could be disastrously counterproductive to give a high dose of psychedelics to someone before dying if they are not prepared and in agreement," said Dr. Richards, "because they are holding on to their body as much as they possibly can."

Because psychedelic mushrooms remain a Schedule I drug, there are no currently accepted uses in palliative or hospice care. Most hospice patients are funded by Medicare, which does not allow payment for an illegal substance to treat any symptom of distress. Nor can a nurse working for a hospice provider legally administer a Schedule I drug. The patient must be able to self-administer the mushroom if they can even get it. And no insurance company is going to insure a hospice facility that administers Schedule I drugs. "In America today, if an RN administered a psychedelic mushroom product to a hospice patient, it could very likely end in the loss of their license and probable incarceration," said Roux. Until the day when states offer dual licensing for, say, RNs and psilocybin service providers, or the federal government specifically reschedules psilocybin (or psilocybin analogs), mushrooms are unlikely to be used by hospice or palliative care programs. There are very high walls built around the use of this substance, said Roux, "regardless of its ability to engender profound and positive changes in one's relationship with death."

There is a point where a patient is likely too far along to benefit from a psychedelic. Roux has worked with several hundred patients at the end of life, "and two to three months out, there is in general a shift in consciousness if the patient is dying from a disease like heart or respiratory failure." (He noticed this shift less in cancer patients and not in those suffering from cognitive decline like Alzheimer's.) "Once you get past that two-month mark," he said, "people are often medicated to manage nausea, pain, or anxiety. All these medications can impair one's psychedelic experience. Three or four weeks away, you might not have the cognitive ability to make meaning out of a transcendental experience."

Horizons: Perspectives on Psychedelics is the longest-running annual psychedelic conference in the world. In 2022, it was held at the Cooper Union's Great Hall in New York City, where some of the country's most important speeches on civil rights have been held, and where Roland Griffiths, the scientist whose work on psilocybin in the early 2000s ushered in the current era of research, made a remarkable appearance via video. We in the audience thought he was going to talk about his work on psilocybin and terminal cancer patients. And he did. But the patient whose journey he described for us was himself. "Over the course of that study, I spent many hours with the participants and often wondered how I would deal," he said. "Now I know."

Dr. Griffiths was a psychopharmacology researcher at Johns Hopkins for more than fifty years. In temperament, he's like the Mr. Rogers of psychedelics: reasoned, ethical, sincere, mild, sweet. He became curious about altered states of consciousness because of his longtime meditation practice, which prompted his first study on psilocybin and mystical experiences. "There is something about these experiences that relates to the nature of human consciousness," he said in a bedtime-story voice. "If you turn your attention inward, you can become aware you are aware. An indisputable and profound inner knowing arises that we can only access individually; it's at the core of our humanity. It feels both precious and true, giving rise to this ethical impulse for mutual caretaking."

In early 2022, Dr. Griffiths underwent a routine colon screening and awoke from anesthesia to learn he had stage IV cancer, which turned out to be largely unresponsive to treatment. Initially, he was in disbelief. "It felt like a dream, that it couldn't be true. But over a few days, I quickly began to explore the range of psychological states that emerged under such conditions: depression, anxiety, fear, resentment, denial, combat—fighting the cancer—all of which seemed uncomfortable and unwise." What he had learned about the "nature of mind" from his meditation practices and from his research into psychedelics became immediately applicable, he told us. And then he dived in. I felt like I was watching him swim, calmly and softly, in the warm pool of his emotions. Here's what he said, somewhat edited.

"The key insight is, we don't need to identify with thoughts and emotions as they arise, but instead, we can turn with great interest to investigate the present moment and we can cultivate gratitude for the astonishing mystery in which we find ourselves. In principle, we can do this at any moment we choose to. We are highly evolved, sentient creatures. For me, the psychological off-ramp from potential emotional misery has been the cultivation of gratitude for the precious mystery of life itself, of being conscious, of being awake to the mystery of this present moment. As unlikely as it may seem, my wife and I have experienced this diagnosis as a gift, as a blessing, really, and I've often reflected what a tragedy it would have been had I been run over by a bus on my way to a mundane medical screening, because today, I am more awake and alive and grateful than I have ever been in my entire life." Roland Griffiths died on October 16, 2023.

Dr. Griffiths's testimony moved everyone at the conference. But all the testimonies are intense and important: the fellow with trich, the grieving moms, the freshman class of MDs who were thrust into the battlefield of COVID, the dying. For them and more, psilocybin is showing potential to mitigate the symptoms of depression, anxiety, and addiction disorders. We diagnose and treat these conditions as distinct and separate, but maybe they aren't all that different. Maybe, as the psychiatrist Christopher Pittenger speculated, these diagnoses can be understood as a "collapse of meaning in one's life."

I actually googled "the meaning of life" (in itself a weird modern-day way of exploring a philosophical question), and up came a thirty-year-old Bill Moyers interview with Joseph Campbell. In it, Campbell said he didn't think it is the meaning of life that we seek. It is the rapture of being alive.

And maybe that is the magic *Psilocybe* mushrooms occasion.

10

Spiritual Trips

A walk with the late mycologist Gary Lincoff in Central Park elicited a steady monologue on the state of favorite trees, the best time of year to eat dandelion greens or any number of weedy things, the underground location of a fungus yet to flush. He kept copious field notes on the flora and fungi he observed on his daily walks. Ralph Waldo Emerson said that spending time in nature is the closest we can come to the divine, an insight, perhaps, founded in the vast natural resources of our country. But Gary didn't have to go into the grand Rockies or virgin Adirondack forests to find communion with nature. He found it along the edges of parking lots, in trampled public parks, in the wells of city trees. He didn't need to see extravagant fungal fruiting bodies to get excited. Hand him a stick and he'd find something growing on it to swoon over. And this was Gary's great gift: he inspired his students to see the divine in the humblest places and in the humblest of organisms.

That's a type of spiritual experience that falls outside the definition of many established religions. Similarly, for some, psychedelics provide an alternative way to experience the divine.

Magic mushrooms can bring about a deep and ecstatic conviction that all living things are connected and all are worthy of love and protection. James Fadiman describes it as an awareness of relativity: "I see that I am a link in an infinite hierarchy of processes and beings, ranging from molecules through bacteria and insects to human beings and maybe angels and gods. From this it is but a short step to the realization that all forms of life and being are simply variations on a single theme."

This sense of unity has also been called *nature relatedness*, and it's one of the varieties of religious experience, according to William Richards. Psychedelics may not only quicken those feelings of connectedness with nature but also the psychological benefits that come with them. In their paper "From Egoism to Ecoism," Hannes Kettner and his associates found that psychedelics increase one's sense of connectedness with nature, leading to decreased anxiety and "a greater perceived meaning in life, higher vitality, higher psychological functioning, greater happiness and positive effect." The mystical experience of nature relatedness is, in effect, a kind of spiritual healing.

The psychedelic-inspired mystical experience, whether connected to nature or not, has motivated all sorts of folks to participate in spiritual practices, to start churches (of varying degrees of churchiness), or to return to the faith of their childhood. In the last twenty years or so, more religious professionals and practitioners have started to think the mushroom might even help them minister to their congregations. It's not a new idea: in the Mazatec tradition, the psychological effects of magic mushrooms cannot be separated from the spiritual. Individual, social, and environmental well-being *is* spiritual well-being.

Magic mushrooms that cause numinous states (that is, a sense of divinity) are known as *entheogens*. The word comes from the Greek *entheos*, meaning "God within," and *gen*, which denotes the action of "becoming," or becoming divine within. The term was coined in 1973 by the classicist Carl A. P. Ruck, during a meeting called by R. Gordon Wasson to come up with a replacement term for *hallucinogens* and *psychedelics*, which the participants felt inadequately spoke to the ontological and potentially religious aspect of a trip. *Entheogens*, they thought, "would be appropriate for describing states of shamanic and ecstatic possession induced by ingestion of mind-altering drugs." The term, according to the author Andy Letcher, was about distinguishing religious use from recreational use. But terms are slippery things. Psilocybin is also called a *medicine* by those using it for therapeutic purposes, and yet indigenous practitioners might say you cannot shear off the drug's therapeutic applications from its spiritual implications, because the entheogenic aspect of a trip is part of the cure.

Spiritual trips, or trips in search of the mystical experience, tend to be large-dose trips. Sometimes very large. In his book *The Psychedelic Explorer's Guide*, Dr. Fadiman quotes a microdoser who said, "My microdose mentor once told me that at the very lowest microdoses you see how much God loves you. If you take a bit more you also see how much you love God and if you take quite a bit more, then of course it gets pretty hard to disentangle exactly who you are and who God is."

I never considered myself a spiritual person—I am not associated with any established faith. Nor had I ever been swept away by a religious service until I attended a *velada*, an evening mushroom ceremony conducted by a Mazatec *curandera* I will call Doña Lucia. A *curandera* (or *curandero*) is a healer. Like shamans elsewhere in the world, they often arise from lineages of healers and serve their community's spiritual, mental, and physical needs. Doña Lucia has become a bridge of sorts between Mazatec culture and the nonindigenous world. Over the years, she has led traditional psychedelic mushroom ceremonies modified for nonindigenous people. That doesn't mean she conforms to any consumerist expectations. It's not like I could pay with Venmo, and up until the last moment, I wasn't sure where I was supposed to be or when. It turned out attending this *velada* was more like harvesting grapes than making a doctor's appointment: you just had to be ready when it was time.

We gathered on a wooded piece of land, in a large, round studio with a wide oculus. I parked next to a half dozen other cars and brought my gear inside—a camping mat, a blanket, a pillow, water, my notebook, and a loaf of homemade bread to go with the "nourishing soup" we'd share afterward. I was looking forward to that, as we had been instructed to fast all day. I found a spot next to a young Mexican couple who had set up a double-size mat. They both wore white. There were several Mexican people there, all wearing white, which a Mexican friend explained was a symbol of purity of spirit, cleanliness, "and in general, shamanic attire in Mexico." At the head of our circle of mats was a table draped in a cloth covered in colorful embroidery, scattered with cacao seeds, and before it on the floor, another cloth, with vases of flowers that people had brought and personal offerings of stones and shells. Behind the table was a tapestry depicting Our Lady of Guadalupe enclosed in a mandorla, an almond-shaped aureole, like a halo around her whole body. In contrast to altars I've seen in the Vatican, in my mother's Episcopalian church, or in synagogues, this one was distinctly feminine.

Altogether, we were about fifteen people of all ages, white and Latinx, from willowy young women to older bonded couples. Some people, having participated in these ceremonies before, embraced, but there was also a kind of anonymity to the group. We smiled at one another, acknowledging without introductions. Doña Lucia had a young assistant—I'll call him Peter—a soft-spoken man with a mass of wavy black hair and drawstring pants. He welcomed us and had us assemble around an outdoor firepit, where we sat on cut logs, warmed by a small fire, pine needles crackling. Up until then, I had not seen Doña Lucia. The person who stepped up to the fire, who commanded all our attention, was a tiny, self-contained, and reserved middle-aged woman, serious but warm, incredibly warm. She wore an embroidered blouse and dangly earrings, her long black hair in a bun. She had a maternal authority that seemed protective; I instinctively felt she was there in my best interest. She smiled, and Peter translated some welcome comments, and then she began murmuring prayers.

Peter gave us each a handful of cedar dust, and while Doña Lucia prayed, switching from Spanish to Mazatec and back again, we made our silent petitions and threw the dust in the fire. I imagine our group was participating for a variety of reasons: curative purposes, mind expansion, spiritual quickening, but this was a Mazatec ceremony—or a version of one—where healing and spirituality are indivisible.

As I threw my handful of cedar dust into the fire, I returned to what I'd been wrestling with all along: how to be good with my stage of life. Peter walked among us, smoking us with incense, praying in Spanish all the while, even when he stepped aside to refill his burner. We sat patiently as Doña Lucia swatted us in turn with a bouquet of mugwort and rue in a pattern, shoulders, forehead, arms, forehead, laps, forehead. Then she swished the plants behind her, as if brushing something off us and away, all the while murmuring, murmuring. It was an intimate yet proscribed act, not unlike receiving Communion, but one perhaps reaching back into a more distant time.

Andy Letcher wrote in his paper "Mad Thoughts on Mushrooms: Discourse and Power in the Study of Psychedelic Consciousness" that a drug becomes an entheogen based on set and setting. Our common mindset was respect, I think, for this woman and her culture. And our setting? Doña Lucia's ceremony was simple, barely adorned, yet it felt as religious and saturated with gravitas as any church service I've ever attended. At the end of the fire ceremony, Peter threw more cedar on the flames and suggested we pray for our loved ones, and for ourselves, and he ushered us back into the studio to take the mushrooms.

Our doses were the standard high dose of 3 grams dried, whole *P. cubensis*, the Golden Teacher strain, and about an hour into the trip, an optional booster of 1–2 grams would be available. The mushrooms were served on a paper plate with a pair of cacao seeds, which were hard, like stale almonds—not so easy to eat. I did my best with the mushrooms, too, but my mouth was quite dry because I'd been avoiding water all day so I wouldn't have to pee at the height of my trip. (Afterward, I learned that mushrooms just make some people need to pee frequently.) I got all the caps down but slipped the stalks into my backpack—when I got home and weighed them, it turned out I had only consumed 2 grams. From the minute we lay down and received our mushrooms, Doña Lucia sang. And chanted. And prayed.

Dr. Fadiman said if a deep spiritual journey is going to unfold, it will almost always begin before the second hour is over. But what is a deep spiritual journey? How does anyone know they've had one? William James (1842–1910), the American philosopher and psychologist, proposed in *The Varieties of Reli-*

gious Experience in 1902* that mystical experiences are central to the idea of religious experiences, and while they might include both pleasure and pain, you can identify a mystical experience by two factors: ineffability and noesis. An ineffable experience is one you can't describe. It's too great, too powerful, too beautiful. Noesis is that particular species of perception aligned with inner wisdom; like a gut feeling, it resonates with truth, what James called an "inner authority" that affirms a sense of something existing beyond our normal comprehension. This feeling of certainty is similar to the personal insights about self and family that trippers may experience. However, the basic message of entheogens, wrote Huston Smith in *Cleansing the Doors of Perception*, is that there is another reality. Or, in the words of my friend Ezra, "My reality isn't the only reality."

Entheogens and the Origin of Religion

The entheogenic trip has inspired many people to think, hope, believe, and argue that mushrooms jump-started religion. It is, of course, very hard to prove what prehistoric people were thinking about magic mushrooms, if they were thinking about them at all. But there is a healthy subculture of mushroom enthusiasts who search for secret or hidden signs of mushroom cults in ancient history, medieval superstition, and the Abrahamic religions. According to Brian Muraresku, author of *The Immortality Key: The Secret History of the Religion with No Name*, an ancient, archetypal religion that involved psychedelic sacraments seeped into

* The Mystical Experience Questionnaire used in studies of psilocybin and mystical experiences was founded on James's work.

the Greek mysteries and eventually early Christianity. He describes these psychedelics as "the original Eucharist" suppressed by institutional Christianity, a "Religion with No Name" that nonetheless continued and was shared by esoterics through signs hidden in plain sight. It's all very *Da Vinci Code.*

The idea first hit the mainstream when, in his 1957 *Life* magazine story, R. Gordon Wasson proposed that ancient humans may have encountered *Psilocybe* mushrooms, consumed them, and imagined God. Once that idea was set loose, the search for the fungal origin of religion took off. Most speculations about ancient mushroom cults and shamanic mushroom practices are based on myth and interpretations of petroglyphs, rock engravings, and sculptures, often deteriorated by time and weather. Since fungus-attributed imagery is not always literal but often abstracted, it's very much open to interpretation. Nonetheless, there is a canon of presumed evidence of mushroom use. The Bee Man of Tassili n'Ajjer, part of an important UNESCO prehistoric rock art site in the Sahara dating from 9000 to 7000 BC, seems to be adorned with little mushrooms. He was first outed as a magic mushroom shaman by the McKenna brothers in 1986. The Selva Pascuala rock paintings (6000–4000 BC) in Villar del Humo, Spain, also sport what could be interpreted as a row of mushrooms near an image of a bull (the implication being that coprophilic mushrooms are growing from bovine dung), as proposed by Brian P. Akers and colleagues in 2011.

Giorgio Samorini, an Italian psychedelics researcher, wrote that in most cases, the mushrooms represented in stone sculptures and reliefs are of the "entheogenic" kind. Why else would the artists be inspired to depict them? These include the "umbrella" or "mushroom stones" of central Kerala, South India (1000 BC–AD 100), and the four-meter-high "mushroom stones" in Thrace (which are naturally formed). Mushroom-shaped stone sculptures in Guatemala may be the real thing, since the entheogenic use of mushrooms is well documented in Mesoamerica, but others are more dubious. Samorini has suggested that what have long been called the axe engravings on megaliths number 4 and number 53 of the Stonehenge monument are actually mushrooms. He also reinterpreted some Bronze Age engravings in Scandinavia depicting figures on ships as carrying mushrooms. Shane Norte claims *Psilocybe* mushrooms have been used for time immemorial by the Payómkawichun (the People of the West). As evidence, he sent me a photo of a mushroom or mushroom-like petroglyph on reservation property that he says dates to Spanish contact.

Imagery in early Christian art has gotten similar mycotreatments. In *The Psychedelic Gospels*, the authors gamely hit churches throughout the European Union, inspecting frescoes and sculptures for far-fetched hints of a secret mushroom cult. Maybe the most persuasive Christian image is a thirteenth-century fresco at Plaincourault Chapel in France that depicts Adam and Eve next to a tree of knowledge that looks like an *Amanita muscaria* that wishes it was a tree. There is a name for all this: *pareidolia*, "the tendency to perceive a specific, often meaningful image in a random or ambiguous visual pattern." It's like seeing Jesus's face in a piece of burnt toast, and about as convincing.*

So far, there is little verifiably recorded entheogenic mushroom use prior to our current era. It is well established, how-

* Most of these mushroom hunts through history focus on *Amanita muscaria* and, to a lesser degree, ergot (*Claviceps purpurea*, a fungal pathogen of grains that can cause LSD-like effects). R. Gordon Wasson spent a lifetime looking for evidence of entheogenic mushroom use in texts, as did Carl A. P. Ruck, Jonathan Ott, John Allegro, and many others, whom the mycologist Britt Bunyard calls, in his book *Amanitas of North America*, "mycological mythmakers."

ever, that *Amanita muscaria* is consumed for sacred and magic purposes by shamans and healers in the northern latitudes, like Siberia, and the Aztec and their descendants, like Doña Lucia, who use *Psilocybe* species similarly.

The Mazatec Way

Psychoactive mushrooms, which are called *Ndi Xijtho* ("little things who sprout") in the Mazatec language, have been and are used in ceremonies in multiple communities and among multiple tribes throughout Mexico, and up and down Central and South America by indigenous, mestizo, and neoshamanic and transpersonal healing communities, and, to a degree, in Nicaragua, Colombia, Bolivia, Peru, and Chile. They are used in ceremony primarily for curative purposes (in contrast to the variety of ways nonindigenous people use them).

The ceremony has changed over the years and across indigenous communities, adaptations "which are legitimate and ongoing," said Ismail Ali, director of policy and advocacy at MAPS, "though different from long-term historical use." But much of what is known by outsiders is the ceremony that was and is practiced in Oaxaca.

In the decades since R. Gordon Wasson's description of his trip with the *sabia* María Sabina, her village, Huautla de Jiménez, has hosted a sacred tourism industry, mainly by neoshamans catering to a variety of tourists. The Mazatec researcher Sarai Piña Alcántara has described the different types of tourists: psychonauts looking to experience the mushroom in the Mazatec tradition; Mexican artisans who come for local festivals and may use the mushrooms; and mycologists who collect specimens, often without any community assent. Pilgrims come to trip or to study to become a neoshaman themselves, like facilitators who plan to work at retreats, and other healers roll into town to offer therapies like massage.

This contact has changed the *velada*, said Mario Alonso Martínez Cordero in an interview. He's a philosopher of religion who studies the Mazatec ceremony. "Most people go to Mexico with an idea of what the *velada* should be, and sometimes the healers adapt to the ideas of the tourists." But the *velada* phenomenon is not stagnant, any more than the Mazatec people are. It is constantly reconfigured, wrote the historian Osiris García Cerqueda, by forces of nature and culture, including by capitalist development in the region. That modernization characterized the ceremony I participated in with Doña Lucia.

The Mazatec ceremony is held at night, in darkness and mostly silence with the petitioner and sometimes with family members as well, and conducted before an altar. The mushrooms and the patient are usually cleansed with incense derived from the resin of the copal tree (*Protium copal*), and then the mushrooms are eaten raw, in pairs. Some *chjines*—the Mazatec word for "healers"—do not eat any mushrooms, or very little, but "the ones that do eat mushrooms usually are the ones who are more authentic and deeper," said Cordero. Prayers are offered on behalf of the patient, some to Christ, but they may appeal to saints of all types, including Mazatec divinities, depending on the patient's needs: for a woman having trouble getting pregnant, for example, the *chjine* might pray to the Virgin Mary.

The psychologist Rosalía Acosta López, who is based in Huautla, explained the *chjine* empathizes with the patient "and manages to see and feel what the patient sees and feels. The man or woman of knowledge, already under the effect of the mushroom, sings, prays, and allows the patient to release what they need to in that moment, whether it be through singing, laughing, or crying. They know when to intervene, to pause and focus their care at a time when the patient is reliving a traumatic moment or perhaps feels physical pain. The *chjine* will ask God or the *chikones* [supernatural beings] to relieve them." After about five hours, when the psilocybin has worn off, the candles are relit, and prayers are resumed. The patients may be instructed to take an offering to the chapel or sacred place of a particular saint or divinity or consume a medicinal tea. A restricted diet or sexual abstinence may precede or follow the ceremony, and sometimes a purification ritual will precede or follow, too. Cordero described the practice of passing an egg over the patient's body, which some healers believe absorbs body energy, after which the egg is discarded.

From start to finish, the traditional ceremony is imbued with Mazatec cosmology. "When the Mazatec say *God*, they say *Father, Mother, God*, or *Father, Mother, Nature, God*," explained Cordero. That is because the Mazatec believe in the beginning there was no difference between male and female and nature. Duality, the separateness of things, happened as the world became manifest. But the mushrooms take us to the time before the manifestation of separate identities, an important place for the patient to go, because for the Mazatec, "we are interconnected with everything else and when you lose sight of that," said Cordero, "illness can manifest." I take from this that healing, in the Mazatec tradition, is not just individual, it is relational.

Acknowledging ancestors and recognizing the individual as part of a bigger whole is key to healing, said Cordero at the

Horizons Northwest conference in 2022. "We are not isolated entities. Everything we do affects others. Illness is an imbalance in relation with otherness, and healing happens when we reciprocate and pay our debts to our community. We are interwoven in divine unity, and we are all a community." The question, he continued, is how the nonindigenous world can ally with the indigenous world, how reparations for their discoveries and acknowledgment of their long history with this drug can be equitably shared moving forward. It's not just a matter of what's right, said Cordero. "This question must be prioritized, because if one of us is missing, we are not complete but imbalanced and wounded."

Dusk descended into night as I lay listening and looking about in the candlelit studio, and then Peter snuffed the candles, and it didn't matter if my eyes were open or closed. I felt very safe, even though I didn't know anyone and I didn't know what was going to happen after the trip was over—by my calculations, about 2:00 a.m.—or where I was supposed to go after that, and I was very hungry. I don't know if Doña Lucia entered a trance, as *chjines* may do, in search of entities who may reveal the origins of her patients' illness, but she sang and prayed for what seemed to me the entire trip. Sometimes Peter would take over, but it is Doña Lucia's motherly voice I remember, hanging between spoken word and melody. She didn't interact personally with anyone that I could see, but rather, she managed us collectively.

Empathy can be a key aspect of the mushroom experience, and that was evident in the way the group functioned as a whole. One fellow broke out into sustained anguished moans, and like babies on a plane, once we heard one person crying, we all became upset. It was as if the room started to heat up with our collective distress. Doña Lucia quickly switched gears from praying to a lilting silly summery song that got a few of us giggling. As more people started giggling, the tension in the room began to drop. Even the anguished man laughed briefly, and then we settled back down again. I realized Doña Lucia wasn't leading us; she was herding us away from harm while we did our own work.

My stream of consciousness that night provided numerous insights into my relationships and my work. I had the presence of mind to ask myself—or the mushrooms—*What is really bothering me?* And I was transported to my grandmother's bedroom in Italy. It was nap time, after lunch. I watched her climb into her bed, tiny and smelling of ironed linen napkins, the bedsprings squeaking as she curled up and took her last breath. It seemed a simple act, one we all succumb to. But it's almost cliché: I asked myself why I was uptight about my age, and I got a dream of a peaceful death.

That peak period did not last long, probably because of the low dose I took. I could feel it when I tipped over the edge, and a self-hugging coziness washed over me. I'd had to pee for a while, but now I felt I could get up and carefully make my way to the bathroom. That simple act of peeing was lulling: the tiniest muscles in my gut relaxed, and the sound of water on water made me drowsy. On the way back, however, I got lost. I had no idea where my mat was (note to self: next time I trip in a round, dark space, imagine where I am located based on a clock), and so I crept around, gently tugging the edge of each blanket to see if it was mine.

At about 1:00 a.m., Doña Lucia and Peter wound down their prayers and began to light candles. Most of us sat up and looked at one another, blinking like newborn chicks. A young woman sang a song that seemed so exquisitely lovely I wept. We were asked to say a prayer, aloud if we liked, but I didn't know what to say. Prayer had always seemed like begging God to pay attention to me. In retrospect, though, I think maybe the fact that I was able to ask my brain questions as if it were separate from the part of me that was asking, and get insightful responses, is a kind of prayer answered. The notion of an inner healer that guided my questions and the content that emerged, from a theological perspective, said Naomi Schulz, an ordained United Church of Christ minister and a chaplain, "represents one aspect of the divine within ourselves. So, from within this perspective, your questions and the answers to your questions came from the divine within you, and the divine without you, and the divine in the mushrooms, too."

As the lights came up, we laid out our food on the floor. The nourishing soup never transpired, and so I cut my hunger with bread and nuts and fruit. Many people were eager to talk, but I felt introverted and unwilling to let go of the trip's otherworldly feeling. I slipped outside, sober enough not to worry about wandering off. A new moon illuminated a conclave of large, smooth boulders near the studio entrance, glacial erratics as big as cars. I was quite warm—while I'd been comfortable inside, the air had become thick with breath and sweat and tears, and I wanted to cool off. I opened my arms and hugged the big, chilly rock. A woman came outside to see the moon and cool down, and I invited her to join me, and we clung to the rocks, and then another woman came out and looked at us curiously, and we invited her, too. Before long, there were a half dozen women cooling their torsos and cheeks on the boulders, patting the rocks in gratitude.

I fell asleep after that, to the sound of my fellow trippers eating and talking and singing little songs. When I woke up, the sun was streaming through the oculus. A few people were still sleeping, but most had disappeared. Those of us who spent the night swept up and folded our blankets and awkwardly said goodbye forever. Driving away, I wondered if the *velada*, traditionally a spiritual ceremony, inspired a mystical experience in any of the people with whom I shared that room. Many of us there were not adherents to the religion it arose from. And yet, it's possible. Religious feelings are practically universal, almost as if they are part of who we are and what our brains can do.

The Science of Mystical Experiences

Neurotheology looks for a neural explanation for religious feelings. That has led researchers in all kinds of directions, from contraptions like the God helmet, which was supposed to elicit a numinous encounter (it doesn't), to symptoms of mental illness, like hyperreligiosity, when a person's religious beliefs, sometimes even hallucinations, interfere with their daily lives. I'm thinking Joan of Arc. Somewhere in between is a millennium's worth of observations by indigenous healers and twentieth- and twenty-first-century studies on psychedelics and mystical experiences. In many ways, psychedelics bring religion and science together.

In 1962, Walter Pahnke, a graduate student in Harvard Divinity School, under the supervision of Timothy Leary and the Harvard Psilocybin Project, conducted the Marsh Chapel experiment (also called the Good Friday experiment), in which twenty seminary students and scholars received either 30 milligrams of synthetic psilocybin or a placebo of 200 milligrams of nicotinic acid to study the reliability of psilocybin to occasion mystical experiences. The setting was a Good Friday service in a private chapel. Unsurprisingly, the blind was broken: it was clear who was tripping and who was not. But 30–40 percent of those students who took psilocybin reported a mystical experience that they still acknowledged six months later. Several participants experienced acute anxiety, and one had to be sedated after he ran out of the chapel declaring the return of the Messiah, which makes one wonder where the line between a psychotic episode and a mystical experience falls. In a twenty-five-year follow-up to the Good Friday experiment, Rick Doblin of MAPS contacted sixteen of the twenty original participants and found little change in their attitudes toward their mystical experience. (The fellow who was sedated refused to participate in the follow-up study and threatened to sue if he was included.)

In the ensuing years, underground practitioners have continued using psychedelics to enhance their spiritual lives or help others enhance theirs, as the Mazatec and other Mesoamerican groups always have. Scientific research into the mystical applications of psilocybin came aboveground with the 2006 publication of a blockbuster paper, "Psilocybin Can Occasion Mystical-Type Experiences Having Substantial and Sustained Personal Meaning and Spiritual Significance." The study was authored by superstars in their respective fields: the addiction researcher Roland Griffiths and Una McCann, a psychiatric specialist, both at Johns Hopkins University. Also credited are Robert Jesse, convenor of the Council on Spiritual Practices (which, since 1993, has been "dedicated to making direct experience of the sacred more available to more people") and a lead writer on a Supreme Court amicus brief that decided a successful religious liberty case,* and William Richards, the Gandalf of the psychedelic scene. The results were striking: most participants reported mystical experiences with positive and lasting personal benefits, as measured by the Mystical Experience Questionnaire.

Since that seminal paper, there have been others, from both the US and abroad, probing the phenomena of a drug that initiates spiritual feelings. One study on the relationship between spiritual experiences and dose size found that the higher the dose, the likelier the participants would have a mystical experience, suggesting a mystical experience is dose-dependent and more predictable at about 3 dried grams (and up). The obvious question is, are religious feelings, in particular the mystical experience, stimulated by excited serotonin receptors? The researchers Frederick Barrett and Roland Griffiths speculated that psilocybin's seeming ability to decrease self-referential processing may be key to the experience of universal interconnectedness—a feeling that is central to the mystical experience.

But this doesn't necessarily translate into an increased dedication to one's established faith. "Drugs appear to induce religious experiences," wrote Huston Smith in *Cleansing the Doors of Perception*. "It's less evident that they can produce religious lives." Judging by the activities of some participants in the Johns Hopkins study "Effects of Psilocybin-Facilitated Experience on the Psychology and Effectiveness of Professional Leaders in Religion," psilocybin seems to have not only reinforced their faith but has given them a new tool with which to augment their ministry. The study included religious professionals from an ar-

* The decision allowed a Brazilian church practicing in the US to use ayahuasca in their ceremonies. It is the standard by which all other churches that offer psychedelics test their luck.

ray of faiths who took two high-dose trips and afterward filled out the Mystical Experience Questionnaire.* The maybe unintended consequence is that the study seeded the psychedelic world with clergy who see the potentiality and the pitfalls of psychedelics in established faith practice.

Can Mushrooms Resuscitate Religion?

Midway through his first psilocybin session at Johns Hopkins, the Episcopal priest Hunt Priest (his real name) experienced a sensation of "something lodged inside me. I imagined it to be energy that had to be released, but with no easy way out. The current of energy, subtle at first, formed in my pelvis and intensified as it moved up my spinal column. It eventually became lodged near my larynx; a blockage was created, which then expanded and began to feel impenetrable. The pressure was so intense that at one point I thought the skin around my Adam's apple was going to blow open. In my mind, I struggled to break it up and quite unexpectedly (and uncharacteristically) began to speak in tongues, the spiritual gift mentioned by Paul in Corinthians. That's not something I ever imagined myself doing." For Priest, who has the merry tenderness of a pediatrician, the experience was a Pentecostal one and profound enough for him to want to see others have the same opportunity. That's why he founded the nonprofit Ligare, an organization dedicated to bringing psychedelic experiences to the Christian contempla-

* The questions ask participants to rate aspects of mysticism (internal unity, or the experience of oneness in relation to an inner world; external unity, the awareness of the life or living presence in all things; a noetic quality—that intuitive sense that something is true; and sacredness, a sense of being at a spiritual height), as well as positive mood, transcendence of time and space, and ineffability.

tive tradition through conversation with religious professionals of all backgrounds.

Trips that fulfill expectations can be powerful things. So can trips that *don't* fulfill expectations. Of the two trips he took as part of the Johns Hopkins study, Rabbi Zac Kamenetz's first was moving, "a next realm, next gate, next portal" trance state, but the second trip was static and empty, what he described as "cosmic boredom." His observational and analytic mind was "very present, so I was able to ask, 'Why is [the trip] happening like this?' But what came up were things like 'Why do I have an issue with paying parking tickets on time?'" While the rabbi might have dived into this potentially rich therapeutic pool, he could not, because he was expecting a mystical experience. For months afterward, he experienced frustration and increasing unsteadiness as he tried to integrate his two very different experiences. "There is this idea that a mystical experience has a set number of qualities. But there were things I experienced that were not on the questionnaire but are mystical experiences in my tradition," said Kamenetz in a conversation. "Where was the question about confronting the brokenness of reality?" That's what motivated him to found Shefa (Hebrew for "flow" or "abundance"), an organization dedicated to Jewish psychedelic support.

When an individual embarks on a trip, he doesn't know what kind of experiences he will have. "There are, of course, numerable drug experiences that have no religious features," wrote Huston Smith. "They can be sensual as readily as spiritual, trivial as readily as transforming, capricious as readily as sacramental." Accepting the unpredictability of a trip is a first step. If you hope to have a mystical experience but don't, and more psychological content emerges, then maybe post-trip psychological therapy might be more appropriate than religious guidance or interpretation. But if an experience feels mystical, then a religious professional may help you understand it from a spiritual perspective. "Science has to deal with us because people [may] have mystical experiences," said Priest in a Starr King School for the Ministry interview, "and science can learn from us, because we have thousands of years of practice in dealing with mystical experiences."

That's why a growing number of religious training institutions are rolling out programs in psychedelic chaplaincy. Trained religious leaders might use psychedelics to enhance their own practice, either individually or at psychedelic retreats expressly designed for clergy, or they might use psychedelics to augment their usual work helping their congregants deal with medical diagnoses, divorce, deaths, and crises of faith. Maybe magic mushrooms could help congregants experience a mystical revelation that brings them closer to God and the bosom of the church. They may even help with harms *caused* by religious views. Queer and trans Christians who are ostracized by the churches they were raised in may suffer from trauma, including the fear that they might not be lovable even by God, said the minister and counselor Tamara Lebak, founder of the Restorative Justice Institute of Oklahoma. For them, psychedelics may open the door to self-healing.

How psychedelics might fit into the superstructure of a particular faith depends a lot on a faith's traditions and where they lie on the spectrum between progressive and conservative. On the conservative end, it's hard to imagine the pope blessing psychedelic ceremonies within the Roman Catholic liturgy anytime soon. But there may be more flexibility about psychedelics in the Jewish tradition. The question of whether magic mushrooms are okay to use would be decided by a *posek*, someone who answers novel questions that didn't come up in the fifteenth century, like whether meat that was grown in a lab is kosher. A posek may address theological questions like whether achieving a mystical state—which has a long history in Jewish tradition—with a drug produces a legitimate mystical experience. (That same concern arises in Christian theology. The Calvinist take on it is you must earn everything you get—so no shortcuts to encountering God allowed. That's ultimately a puritanical ethic, not necessarily classic theology.) Synagogues are autonomous; they are for the most part independent community organizations that don't have to answer to a higher authority. So any particular group may determine if and how they want to use psychedelics.

Quakers function similarly, and that may account for a more liberal attitude toward psychedelics. I spoke with John Custer, who has been a Quaker his entire life (he was eighty-two at the time of this writing). Custer read the 2006 Johns Hopkins study on mystical experiences, and in proper Quaker style, he decided to find out about psilocybin for himself. He attended a retreat in the Netherlands, where, after two underwhelming trips, his third was a full-on mystical experience. "I was bathed in love, the strongest I ever felt. I was in the presence of the divine—not looking at it: I was in union with God and all life and just about everything. That's when the love hit. And at the end, I felt I had a calling from God to spread the word among the Quaker community that their experience of the divine could be radically enhanced." And indeed, with a like-minded colleague, Paul Indorf, he has started an organization they call NuCovenant to explain entheogens to their community.

Quakerism began in England in the 1640s as a response to

the Church of England's authority over individual spirituality. Its founder, George Fox, argued God is in everyone, which was considered heresy at the time. "Quakers have always had a clear and insistent sense that we can have a direct experience of God without the usual trappings of organized religion," explained Indorf. "I often lightheartedly declare that if you're disenchanted with organized religion, check out the Quakers . . . we're a totally *dis*organized religion!" There is no Quaker super authority to prohibit mushroom-assisted mysticism. When confronted with another approach to spirituality, the Friends (which is what Quakers call one another) are supposed to be open and curious; they discuss until they come to consensus (which can take a while). "I foresee discussion about psychedelics as reliable producers of mystical experience," said Indorf, "and I anticipate Quaker groups doing integration work and even organizing retreats in legal venues." It's possible entheogens could become acceptable practice among Quaker and Jewish groups because the structure of their faiths potentially allows it, and because they are established religions, they may have more legitimacy in the eyes of the law.

Less well-known churches that utilize psychedelics have emerged and faded over the decades, and they represent a range of views about what constitutes spirituality and what the government considers a church. As Timothy Leary intoned, "Start your own church!" And they did. For example, some churches feel more like artistic endeavors, like the Neo-American Church, whose primary text was the absurdist Boo Hoo Bible, which employed comics and rantings of various sorts. (They were denied a religious exemption by the DEA.) The Chapel of Sacred Mirrors in upstate New York *is* a work of art, created by the artists Alex and Allyson Grey as a "sanctuary for spiritual renewal through contemplation of transformative [and I would add, seriously psychedelic] art."

Others tap Native American roots, like the Church of the People for Creator and Mother Earth, based on the La Jolla Reservation in Southern California. Shane Norte conducts the Wamkish ceremonies for the community, replacing the traditional hallucinogenic plant datura with high doses of wood-loving *Psilocybe*. Some churches combine psychedelic-assisted revelation with new age spirituality, where practitioners create their own spiritual stew from a multitude of religious and pseudo-religious disciplines, incorporating everything from astrology to megalithic geometry.

Most psychedelic churches don't seem to produce structured religious communities, though arguably the Farm, a spiritual commune established in Tennessee in 1971, has retained enough structure to stay in business, if operating today mostly as a hippie retirement community. Psanctuary is another; a nondenominational faith-based community in Kentucky that centers on using magic mushrooms. In the church's estimation, a microdose is a "subtle sacrament encounter" and a full trip is a "congregational communion." They conduct retreats for their one hundred local and three-hundred-plus international members, even administering end-of-life sacraments.

The founder, Eric Osborne, has been involved with mushrooms a long time. In 2013, he and his wife opened the first psychedelic retreat, MycoMeditations, in Jamaica, adding more partners in later years. Eventually, Osborne parted ways with his partners (but not his wife) because they wanted to push MycoMeditations in a clinical direction, providing more traditional psychedelic therapy, and Osborne was more interested in exploring psychedelic-assisted spirituality. I met Osborne at the

Psychedelic Science conference in 2023. An athletic-looking man, he was wearing shorts with socks depicting green alien heads. "My belief," he drawled in his Southern accent, "is all mental health benefits stem from spiritual healing." Osborne is among those who believe psychedelics were used as early Christian sacraments and that the use of psychedelics today is simply modernizing religion by bringing back the practice of autonomous spirituality.

Another well-known psychedelic church is the Sacred Garden Community. They hold bimonthly ceremonies in the Portal, a community center in Berkeley, California, funded by the Bronner family (of Dr. Bronner's soap and major financial supporters of psychedelic advocacy). To join the church, folks must participate in the community for three months and attend eight events before they are vetted for health and safety, said Livi Joy, an "ordained facilitator" of the church, and, I imagine, compatibility. Only then can they participate in the "sacraments."* The Divine Assembly was founded in Utah by a former Mormon as a spiritual alternative to the excesses of the Latter-day Saints. "There are so many ex-Mormons," said Suzy Baker, cofounder of New York's psychedelic library, the Athenaeum. "And we all bond on the fact that Mormonism can be traumatizing." I've encountered other such refugees from established religions who find a more forgiving spiritual home in psychedelic churches and societies.

What these churches seem to have in common is the belief that with the help of psychedelics, the individual can commune with the divine independent of ecclesiastical management.

* The Decriminalize Nature organization, which works on decriminalization and expanding access to entheogenic plants and fungi, grew out of the Sacred Garden Community.

What they risk, in my view, is falling into cultish practices. What they offer, potentially, is mysticism without prescriptive dogma. None of the psychedelic churches I researched want to be seen as a beard for using drugs. Many churches avoid any public-facing presence, for obvious reasons, and some have joined a young self-regulating body called Sacred Plant Alliance (SPA), an association whose members seek to address the significant challenges of using psychedelics: to wit, how to reduce the potential for harm and how to advance their legal protections under the Religious Freedom Restoration Act. Allison Hoots, the SPA's president and lead author of *Guide to RFRA and Best Practices for Psychedelic Plant Medicine Churches*, explained that their goal is to demonstrate that psychedelic churches can be safe. In her opinion, psychedelic churches offer a more inclusive way to access spirituality and divinity. "But at SPA, we are just trying to get ahead of what a church using these substances should look like, how to reduce inherent risks, and to keep educating the public about sincere, responsible spiritual use."

Psychedelic Churches and the Law

Running a psychedelic church is a risky business, though decriminalization efforts throughout the country are underway, and psilocybin has generally become more socially acceptable. I paid a visit to Zide Door, part of the Church of Ambrosia, in Oakland, California, where psilocybin has been decriminalized. To become a member, I had to sign a document stating my acceptance of entheogenic plants (including cannabis and mushrooms) as part of my religion and pay a small fee. Members get occasional lectures on the entheogenic uses of magic mushrooms in a funky chapel with blue-velvet-cushioned pews and access to a salesroom where mushroom products are sold. Sunday services are conducted by the founder and pastor, Dave Hodges. A sign on the pulpit reads, "Free joints for all during the sermon (everyone must be smoking)," an effort, I think, to break federal agents or police officers' cover. The salesroom feels a lot like a cannabis dispensary: no cell phones allowed, and space between parishioners is strictly enforced. A long glass cabinet holds different cultivars of *P. cubensis* for sale in capsules, some straight, some mixed with functional mushrooms like lion's mane (*Hericium erinaceus*) in a variety of doses. They also sell ten different cultivars of whole dried *Psilocybe* mushrooms and chocolates containing magic mushroom powder. Legally, Zide Door takes the position that it functions under freedom-of-religion laws.

In response to an anonymous complaint, the church was raided in August 2020 and its cannabis, mushrooms, and cash confiscated, though there were no arrests or charges. Similarly, Benjamin Gorelick, "the mushroom rabbi" and head of the Sacred Tribe, a Denver-based psychedelic synagogue of around 270 people, was arrested on felony charges in February 2022. However, by the end of the year, the Denver DA dropped the charges, citing the approval of Proposition 122, which legalized psilocybin and several other psychedelics in Colorado.

Psychedelic churches are hard to prosecute; the Religious Freedom Restoration Act (RFRA) prevents the government from interfering with sincere religious activity unless it can prove a compelling interest to do otherwise.* What constitutes

* In 1993, Congress passed the RFRA in response to a Supreme Court case involving the use of peyote by indigenous Americans in their religious ceremonies. Few psychedelic churches rise to that level of tradition and practice.

a sincere religious exercise, however, is somewhat in dispute. Growing pot for "the good of mankind," for example, is not a sincere religious exercise in the eyes of the law. A few years after the RFRA was passed, the defendant in *United States v. Meyers* argued his pot bust was illegal under the RFRA because his religion commanded him to use, grow, and distribute marijuana. The Tenth Circuit Court didn't buy it and went on to cite twelve factors necessary to meet the definition of a church, among them metaphysical beliefs, important writings, structure of organization, and holidays. As one student of theology at Harvard in 1993 observed, if a church is going to include entheogens, it needs to have an entheology.

A church must avoid a compelling reason to be busted, like threatening the health and safety of the public (so, no human sacrifices), or not securing their mushrooms and thereby allowing neighborhood kids to raid the tabernacle. But outside of responding to neighborhood complaints, busting psychedelic churches isn't a high priority for law enforcement. Drug cartels aren't really into magic mushrooms, and trippers are rarely associated with violent crime.

Churches can apply to the DEA for an exemption, though it is a notoriously poky process, and the DEA has its own criteria for what constitutes a church. It's not surprising, then, that numerous church leaders I spoke with said their plan was to wait for the DEA or state authorities to make a move, if any was to be made at all. "I am not an outlaw," the founder of Psanctuary told me. "I have long held the belief that no plant is illegal. I've always behaved as if no plant is illegal. We have been instructed the burden of proof is on the government. We put the DEA on notice this is our constitutional right. That's how we roll."

Psilocybin can occasion mystical experiences, more reliably

perhaps than a church service, and those experiences can reaffirm and sometimes redefine faith. Considering the US's rapid secularization (30 percent of us are religiously unaffiliated), some religious professionals are eager to embrace psychedelics if they can bring more people back into the fold. On the other hand, I have met many folks who trip in small self-organized groups—something like the Buddhist sangha—which they say enhances their spiritual lives. For them, a spiritual practice with mushrooms doesn't need a church at all.

11

Psychedelic Community

Ever since psychedelics escaped Albert Hofmann's lab at Sandoz in the 1960s, various tripping collectives have flourished, from the Grateful Dead family and the Living Theatre to contemporary organizations like the Asian Psychedelic Collective, which is involved in culturally responsive harm reduction and education. J. Christian Greer, an expert on psychedelic collectives, suspects there are thousands in the US today, groups that have "placed the collective use of psychedelics at the core of their newly invented religions, communal experiments, and activist collectives," he said in a presentation at the Ecological Spiritualities Conference in 2022. "Groups, not individuals," said Greer, "represent the real engine of the consciousness revolution over the last seven decades."

Tripping in groups runs the gamut from the joyful hedonism of a rave to the quiet camaraderie of cancer patients. From a clinical perspective, group trips with integration afterward can produce excellent outcomes and are cheaper than private therapy. From a social perspective, they can lead to memorable experiences, long-term bonds, and personal fulfillment, and for identity-based groups, tripping or integrating with like-experienced people can help address harms. Any singular trip is a private affair, but in community, the positive outcomes can be multiplied.*

Clinical Trials and Group Therapy

Most of the clinical trials being conducted in psilocybin-assisted therapy are individual, but the idea that group psychedelic-assisted therapy might be a scalable model is something that interests researchers like Brian Richards. In a study, he and his colleagues explored psilocybin and group therapy for cancer patients. The drug was administered individually, but the patients received wraparound therapy together in a community cancer center. The researchers found clinically meaningful, rapid, and sustained improvement in depression symptoms in these patients. Those involved said the group format made a difference, and many continued to meet as a group (on Zoom) once a month for at least another two years.

The social connectedness patients feel in a group "may be a fundamental, underlying mechanism to therapeutic change," wrote Alexander Trope and his colleagues in the paper "Psychedelic-Assisted Group Therapy: A Systematic Review." Another study that looked at how psychedelics might function in a group therapy format for male AIDS survivors with demoralization issues similarly found that participants felt that the positive emotions and social connectedness they experienced during the trip were reinforced in group conversation afterward. Indeed, in individual psychedelic therapy trials, participants have asked to meet other

* So can negative outcomes. Some psychedelic groups have become cultish and dangerous. The infamous Manson family is an example.

people who have been in those trials, in part to corroborate their experience.

Another benefit of group psychedelic therapy is cost. In the future, should psilocybin therapy become available, group-based therapy would be cheaper than private sessions. It already is at Oregon's Eugene Psychedelic Integrative Center, which offers group pricing. The one-on-one trip model is expensive; a prescription of one to three trips could entail forty-plus hours of clinical labor costing thousands of dollars, and prices might climb even higher if, as some institutions have suggested, *two* facilitators or therapists are required to avoid the possibility of abuse.

Psychedelic group therapy could achieve the same things sober group therapy does: give participants the opportunity to explore their feelings in a safe environment and find community with like-minded others. Bad trips might be addressed and ultimately produce the same positive outcomes a tripper might experience in one-on-one therapy. And there's a certain protection in numbers: the kind of manipulative abuse that has been reported between some psychedelic therapists and vulnerable trippers is less likely in a group setting, though there is always the possibility of feeling marginalized in a group, or the opposite, finding yourself sucked into a cultlike environment. But the overall value of the group aspect of psychedelic therapy is not lost on the many identity-based psychedelic groups that exist today. For them, a group integration experience can offer contextualized support.

Identity-Based Psychedelic Collectives

Public-facing psychedelic groups form for purposes of education and sometimes advocacy, but predominantly for post-trip integration therapy. Identity-based integration groups are important because they can provide a safe place to discuss concerns of all sorts, of course, but particularly those faced by marginalized people. For example, people of color inevitably deal with issues of race during psychedelic journeys, said NiCole T. Buchanan, a clinical-community psychologist, in the video "Chacruna Debunks 6 Racist Myths from the Psychedelic Community." Groups like the Arab Psychedelic Society, an online group whose founder, the psychedelic researcher Haya Al-Hejailan, lives in Saudi Arabia, do not advertise the use of psychedelics, but they do offer integration sessions sensitive to the Arab experience. The same is true of the Asian Psychedelic Collective (APC). It was created in part so members can find the support and community they need. "We *all* deserve culturally responsive care," pointed out Preeti Simran Sethi, founder of the APC. Numerous groups serve Black communities, like Black People Trip, which programs educational workshops and connects people to BIPOC psychedelic therapists.

Gender-oriented groups, like the Psychedelic Liberation Collective, host integration sessions for like-experienced people, mainly queer and BIPOC folks; Queer Medicine Community can help folks find psychedelic services and answer questions like "Is it okay to do psychedelics while on hormone therapy?" Women's groups, like the Psychedelic Sisterhood, an education and integration group, allow for difficult conversations, too. "I started the sisterhood because we needed a group that could speak plainly," said founder Pamela Jackson. "Women might have to overexplain their feelings to men and may not want to talk about things like menstruation with a bunch of guys anyway." Marginalized people often experience intersectionality, where social categorizations such as gender, race, class—and subsequent harms—may apply. That's why over the last few years, Jackson has cultivated more women of color and folks who identify as women.

Beyond gender and race, a syndrome may bring together a group. Over many months, I sat in on the Autistic Psychedelic Community's Sunday afternoon Zoom conversations, "a group dedicated to peer support advocacy and education at the intersection of psychedelics and neurodivergence." The group was cofounded by Aaron Paul Orsini, who has since become a—if not the—leading advocate for psychedelics and autism. They taught me a lot about what autistic people go through on a day-to-day basis, and for that insight, I owe a great deal to the generosity and openness of the folks who let me listen in. The groups averaged about fifteen people from all over the world; the rules were simple: movement and stimming (that's repetitive sounds or movements like rocking) are okay anytime, as is text-based chat. And those chat boards were active! People wrote long comments with fully developed ideas so quickly that I couldn't keep up and had to take screenshots.

The group I observed changed weekly and was consistently gender diverse, though dominated by white, Asian, and Middle Eastern participants. A lot of people were tuning in from their parents' homes, some not having gone out in weeks. One woman logged on from her bedroom in Sweden, shrouded in a blanket. She explained how difficult it was for her to go to the grocery store. Her senses could rapidly become overloaded by cereal choices, fluorescent lighting, and beeping noises. During one session, she recounted a meltdown where she just crumpled in the store. The grocery clerks didn't know what to do and called the police, who "shined flashlights in my face," causing her to become even more distressed. What she needed, she said, was for someone to calmly ask if she was autistic and then just stay by her for the short time it would take her to recover. It's a simple human gesture that most of us wouldn't know to make, but it reminds me of being with someone who is having a bad trip. You can't talk them down, and calling the police might exacerbate matters; the best thing to do is just be a calming presence until their anxiety wanes. In fact, some folks in the group said they will call a psychedelic emergency hotline—even if they are not tripping—for reassuring words when they are having a meltdown.

I learned a lot about what life is like for neurodivergent people from these sessions. In general, folks in the group benefit from psychedelics the same way neurotypical people do. They don't take psychedelics to change the way their minds work. None of them wanted to stop being autistic, because even though "being in touch with and attuned to every feeling and energy of the world is awful and exhausting," it's also beautiful. They take psychedelics for a range of reasons, like increasing their joy of being around other people. This is important because neurodivergent folks can be reclusive or avoidant. Some people in the group said psychedelics enabled them to ask for help. Others said they can handle sensory inputs better after tripping. A person I will call Birdie explained that prior to psychedelic use, they shied away from touch and connection. "Psychedelics facilitated connection for me," they said.

While there are people on the spectrum who might benefit from psychedelics and other interventions, autism is not a disease. If you are autistic, you will be your whole life. Subsequently, the goal for many neurodivergent people is to find a way to be autistic in our society. "Living in an accepting culture would be better than drugging ourselves to be more accessible," pointed out one Finnish member. Indeed, I heard Orsini speak at the Psychedelic Science 2023 conference, on a panel called Affirmative Care for Trans, Queer, and Neurodivergent Communities. When, at the end of the program, it was time for him to share some concluding remarks, he said that, "in the spirit of authenticity," he would speak in the way that came naturally to him. And then he let loose, in full sentences and paragraphs, speaking so fast I could hardly understand him. It was a beautiful display of his unique mind.

You could describe these affinity groups as social justice groups within the psychedelic world, where psychedelic and activist activities overlap. What links the psychedelic and social justice movements is medicine, said the author and psychedelic advocate Nicholas Powers, "because as long as the medical model is the dominant frame, then you will have social justice activists who will say we also are focused on the pain caused by society itself." Indeed, public-facing BIPOC integration groups often go beyond personal trip integration and education and advocate for greater presence in clinical trials, practitioner training, and leadership positions in the field.

Integration sessions for specific demographics can also be found embedded in regional psychedelic societies; there may well be one in your state. For example, the Baltimore Psychedelic Society offers integration circles for the community in general, one just for men, and another for trans and nonbinary people. The function of these societies can vary—some are portals for fee-based psychedelic services—but their purpose is consistent: to serve the psychedelic needs of their community. And that's the mission of the seminal psychedelic community, the people of Huautla de Jiménez in Mexico. They preserved the *velada* of their ancestors, which plays a role in preserving community. And preserving community, pointed out Mario Alonso Martínez Cordero in a Horizons talk in 2022, is key to the healing process.

To find a community organization, you can check the Global Psychedelic Society (globalpsychedelic.org) or Psychedelic Experience (psychedelicexperience.net). You can also check the directory at Tripsitters (tripsitters.org), or social media outlets like Instagram or Meetup (meetup.com). Other groups may take a little digging, but they are out there.

Private Tripping Groups

In contrast to integration groups, you aren't going to find a psychedelic collective that trips together on the internet. Or if you do, be wary. Entry into tripping groups is by word of mouth and by recommendation and referral, not online recruitment. The underground nature of these events, the illegality of their actions, and the subsequent problems with involving medical or police personnel if something goes wrong require members to carefully screen the new folks who join. That vetting process does more than protect the integrity of the group: it may also protect those who should avoid psychedelic drugs in the first place. An underground community has a vested interest in knowing their members well.

Private tripping groups come in all shapes and sizes, but often, they last for years, and the members become like a family. And they abide because, as my friend the yogi Pasha Gurov, whom I met at Blue Portal, said, "You need other people to know yourself." A case in point: a Brooklyn-based men's group of six or so professionals meets two to four times a year in a secluded Airbnb in the woods of upstate New York. They trip with different psychedelics, cook together, hike in nature, and conduct group intention and integration sessions. "We each realized we'd been raised to compete with other men and not to support each other," one of the members, a psychotherapist, told me. "So, we're nonhierarchical, there's no leader in our group; all chores, costs, responsibilities, and decisions are shared equally. It's been uniquely healing to speak and to listen within a group of men, to be able to lower your guard in a safe, free space. The psychedelics help lower the walls, but it's the trust that develops in the communal relating that keeps them down."

Private tripping groups may start small, but they can, over time, expand by merging with other groups or by increasing their numbers. A somatic practitioner and psychedelic educator in New York City named Soma explained to me that in Black and brown communities, small private groups use psychedelics to address issues of all sorts, including the personal fallout of endemic racism. She described them as modeled on the homeschooling pods that formed during the pandemic, mainly in cities where psilocybin is decriminalized. Soma is a dignified woman who spoke with the kind of delicacy and caution that seems inevitable given the realities of BIPOC psychedelic use. "There have been a lot of different pods doing this work. In the past, they insulated themselves out of fear of overpolicing," she told me, "so they have kept their own networks secret." But she explained the movement is slowly growing, and the once disparate psychedelic groups are beginning to cooperate on projects like organizing psychedelic retreats. "A lot of us have been doing this work a long time," she said. "My hope is that we do connect more, because we can only become strong and more resourceful together."

I imagine these groups embody a wealth of anecdotal data about types of harm that occur and how psychedelics and integration work helped mitigate those harms. But unfortunately, marginalized people often fear going public, said NiCole T. Buchanan, due to a history of punitive policing and unequal application of the law. It doesn't help that many nonmarginalized folks believe ingrained social inequity can be eradicated if everyone just took psychedelics.

Can Tripping Save the World from Itself?

The idea has long preoccupied psychedelic partisans. In 1970, Grace Slick, the lead singer of Jefferson Airplane, got invited to a party—under her maiden name, Grace Wing—at the White House by Tricia Nixon. Slick said she considered dosing Richard Nixon with a psychedelic to give him a new perspective on the world, but never made it inside, probably because she brought anarchist and anti-war organizer Abbie Hoffman as a date. But the hope that psychedelics can change the minds of the powerful and privileged endures.

If the world's political leaders took a psychedelic trip, would they become more compassionate toward the disadvantaged, more willing to put the common good of the planet ahead of capitalist goals, and less likely to conduct war? That's a conviction held by many in the psychedelic community. Paul Stamets has floated the notion of *Homo ascendus*, that in time, and with ample psychedelic use, our species can evolve to a more enlightened state. Before founding MAPS, Rick Doblin had asked himself, *Why are we prejudiced against different religions, countries, genders, or races?* "You do it through this process of dehumanizing," said Doblin in an interview with *GQ*. "It's all this sense of us and them. So, the core idea was, psychedelics [create] this intuitive, mystical state—you realize we're all connected, and we have more in common than different. Then, politics change. That was the animating vision. The antidote to genocide, to environmental destruction, was the sense of connection. Not just the idea of it, but the felt experience of it."

But psychedelic revelation isn't by nature liberal; different people imagine what a better world looks like differently. Trippers have expressed feeling like the mushroom opened their eyes to conspiracies. The "QAnon shaman," who stormed the US Capitol on January 6, 2020, wearing little more than body paint and a modified Davy Crockett hat, was a psychedelics user. Mushrooms on their own don't skew people left or right politically. A study found psychedelic use is linked to "non-conformist thinking styles," but freethinkers can exist across the political spectrum. It may be that psilocybin trips amplify a mindset that is already present. A trip won't necessarily change who you are. But it might help you see who you are. In the case of the isms, like racism and sexism, change could happen if people choose to use psychedelics with the intention of working on our own biases.

And scaled up, *that* might change the world.

A Place to Trip in Every Neighborhood

So how to scale up? Some folks are thinking about ways trips can reach more people. Journey Space is an online community where members can trip on their own (or achieve an altered state some other way) with live facilitated support. It's teletripping. Cofounder Lewis Kofsky envisions Journey Space as a model for psychedelic-assisted therapy that may be viable not only because of the cost savings and the benefits of group psychedelic work but because telehealth allows for greater access.

Another potential model is the passion of the director of the Brooklyn Psychedelic Society, Colin Pugh, a sweet-natured former stand-up comedian with a habit of tacking "anyhoo" onto his observations. I attended one of the society's Trip Tales events, a regular gathering of folks who talk through their trips. That day, it was held at the Stanford hotel in Lower Manhattan. The hotel had recently outfitted a strip of outdoor café space with yurts, spray-painted Christmas trees, and giant, puffy *Amanita muscaria* dolls. I joined six people in a yurt that smelled strongly of sheepskin, and as we ate Frito pies made with mushrooms instead of meat, Pugh explained his vision of a nationwide network of community-based psychedelic co-ops.

He imagines spaces where people can learn about psychedelics, talk about their experiences, and "get support on their transformation and healing journey" with folks who live in their neighborhood. As psychedelics are legalized, these community hubs could offer guided trips, he said. "The model is community-based healing," he told me, not community in the sense of identity but community in the sense of geography. "I one day hope to form a national organization to help the various psychedelic societies launch co-ops of their own. Anyhoo," he said, "I'm not trying to stop the pharmaceutical model. I'm just trying to imagine a more accessible and hopefully more effective approach." Dennis McKenna agrees. He thinks every community could have "a holistic healing center that includes psychedelics." In which case, maybe someday your neighborhood will have a laundromat, a bakery, and a psychedelic wellness center, too.

Tripping for Fun and Community

Psilocybin can help (or harm) people in need of emotional healing of all sorts. But tripping for diversion, pleasure, and curiosity is a huge part of the psychedelic scene, probably the lion's share. Lots of people trip by themselves, but taking psychedelics with others can lead to a group cohesion that in itself is an exciting experience. People who trip at raves and festivals create temporary bonds that form a kind of transient, ecstatic community. They come together around common interests—music, dancing, drugs—but once their common interest is over, and the band has packed up and left the fairgrounds, the community only exists in memory. That's why every year, when you get off the bus at the Burning Man festival in Nevada, some guy in a tutu offers a hug and says, "Welcome home."

Burning Man was not on my radar before I began research for this book, but it struck me as the ultimate event to attend if I was to understand how group trips create temporary community. Lots of people trip during the weeklong encampment and participate in all manner of social and cultural adventures. That intention—to have fun or blow your mind—may seem immoral compared to the virtuous pursuit of healing. But the beauty of the Burning Man festival is its judgment-free environment. If you want to trip for pleasure, be my guest.

Taking a large dose of magic mushrooms in a social environment like Burning Man or a music festival happens all the time, of course, but that's one of the reasons people have bad trips: it's no fun to disassociate in a crowd. On a lower dose, trippers are likely to feel less anxiety. Many people I met at Burning Man told me they preferred museum, or low, doses for gallivanting around the city. It makes sense: the city *is* like a giant outdoor museum.

The festival takes place in a Pleistocene lake bed, a desert as alkaline and flat as a cake of face powder, ringed by black mountains. Eighty thousand people roll in from all over the world to set up a temporary city of tents and trailers squeezed into seven square miles, much of it open space for art exhibits. I did what everyone does at Burning Man. I tried to take in as much as I could: eat at camps like Waffles and House (*House* as in *house music*—they begin thumping first thing in the morning) and attend lectures like a talk on cryptocurrency and the environment and classes like a psychedelic breathwork workshop, where a woman in a deerskin getup wailed in ecstasy for an hour. (It just made my hands go numb.) I visited the fanciful sculptures and temples adrift in the vast desert like ships on a dried-up sea, and watched the extravaganzas like a drone show, where a three-dimensional, larger-than-life-size whale languorously pursued a school of buzzing minnows across the night sky.

I intended to trip with the rest of that roiling, merry community, but I kept putting if off because just being in the city was stimulating. It was like attending a three-ring circus, a

Mardi Gras parade, and a packed disco all at once. One night, I watched Android Jones's film *Samskara* lying on the floor in the Elephant Camp's projection dome, and it was so intensely trippy I practically wept in gratitude that I was sober. And then there were the dust storms. Bad dust storms, almost daily; a hot, rolling tsunami of dust, many feet high and prickling with its own electricity. The environment was so tough, I just couldn't drink or take drugs. I was sober almost the whole week, more sober than when I am at home. But finally, on burn night at the apex of the festival, I figured it was now or never. I took a museum dose, about 1 dried gram of *P. cubensis*, low enough to be confident I wouldn't become discombobulated and end up at the far end of the playa tangled in the mile-long mesh fences that catch billowing trash.

The burn was impressive. Around its base were rings of fire department personnel in their orange vests and rangers adorned with medals and patches, a thousand-foot perimeter established to dissuade anyone who might attempt a run at the fire. And all around them, the faces of tens of thousands of people glowed and the art cars, a cross between parade floats and sculptures on wheels playing techno music—*boots n cats n boots n cats*—shot off their fire torches and everyone was dancing, and I couldn't tell if I was tripping or if the whole scene was just trippy. The Man exploded in flames, fireworks shot out of his head, and narrow smoke tornadoes slipped from the fire until they ran out of energy and dissipated into the night. Fat yellow cinders hovered in the sky overhead, and I thought they were so beautiful that I couldn't stop taking pictures of them floating and sparkling in the black night.

The next morning, I looked at my photos, and they just looked like cinders. I'm not the only one to take pictures of objects that were fascinating under the influence. Lots of people photograph things like a cabbage leaf only to realize the next day that the image no longer holds the secret of life. But I also took photos of the wide-eyed crowd, faces illuminated by the fire glow, the rows of discarded bikes, fallen over like dominoes, their Christmas lights still blinking, and, unlike my cinders, those images were as wonderful in the morning as they had been the night before. That's when I realized that at Burning Man, communal tripping is not about the trip. It's about the community.

When I returned the next year, soaking rains turned the city into a quagmire of mud, and all the camps were told to shelter in place (the Burning Man Organization has a radio broadcast, "The Voice of the Man"). There was no moving around. People tried and got stuck in the mud for their troubles. Our camp had food and water enough for drinking and cooking, and so we shared what we had with random Burners who were stranded in our party tent like castaways. Indeed, my friend Lili, the mother hen of the camp, my sister, Lisa, and I just kept that kitchen tent open so people could come around for tea or a snack or a Band-Aid, whatever they needed that we had to share. When our campmates would show up to help with food prep, their boots obscured by blobs of mud, we'd gently ask, "Are you tripping?" If they were, we wouldn't let them use the knives. This was communal tripping, too, only I was playing the role of sitter, in gold lamé with paper flowers in my hair.

Raves

People who trip at raves come together into a kind of temporary community, too. The author and educator Nicholas Powers told me about raves he attended in college where the combination of mind-altering substances, loud music, and dancing unified the group, and strangers became family, leading to feelings of positivity and self-worth.

A rave typically starts late at night and continues until dawn. (A young friend invited me to a rave in a Brooklyn park and I thought, *Why not?* And then she told me it started at midnight and went on until 6:00 a.m., and I thought, *Ah, that's why not.*) People dance for hours high on all kinds of psychedelics (alcohol not so much, and the hardest thing to find is a cigarette). "The music is pounding, and I no longer feel I am in my body, and my spirit is carried from body to body, everyone is dancing in unison, hitting this communal, organismic, almost athletic moment," reminisced Powers. "Something is released in all of us at the same time. We stumble from the dance floor to the sidelines. I expand beyond myself into the feeling of others and then am poured back into my body. It's a soul wash, a rite of passage. Raves heal an ingrained sense that my self-worth comes from hustles. It heals a social part of me. That is the potentiating power of a community's embrace." What Powers chronicled has an academic name: "spontaneous *communitas*," the transient experience of togetherness.

Here's how a rave creates community, according to a group of UK-based anthropologists: the combination of rhythmic music, dancing, sleep deprivation, and psychedelic drugs leads to the individual ego dissolution and subsequent group homogenization that Powers spoke of. Status based on money or genetics dissolves. When the group becomes homogenized, it bonds into a community. Throughout history, people have used dancing, drums, and mind-altering substances to create group cohesion (or dancing and drums to alter the mind) and for a variety of purposes: solidarity within a military or hunting party, like the trance dances of the Dobe Ju/'hoansi people, a hunter-gatherer society in Botswana, or religious ecstasy, like the Egyptian Sufi dances, a mystical branch of Islam.

The bonding that happens at a rave, though temporary, is its point. Raves produce many of the same features on a community level that a psilocybin trip accomplishes on a personal level—to varying degrees, ego dissolution and a sense of oneness with the world. In some cases, it may increase the raver's compassion and joy after the party is over. But ultimately, a rave is more about feeling good in the moment—and sharing that feeling with others—than personal transformation. Which is not a criticism. Not everything has to be a therapy session.

Raves are a more contemporary incarnation of what has been going on since psychedelics were first introduced to popular culture. For the last fifty years, a dance party called the Loft, where people dance and trip to strobing lights and intense sound systems, has been going on mainly in New York. Started by a fellow named David Mancuso (1944–2016), these parties have varying locations; there is no advertising, no judgmental doormen. They are by invitation with a fee to defray the organizers' costs. The point of the parties then and now is to create a kind of synchronism among the guests, an ephemeral community. Similarly, house parties, which can last all night or multiple days, started in Chicago in the 1980s, featuring DJs whose music merged disco and hip-hop with early techno.*

* *House music* is so called because the DJs created the tracks at home. Acid house was one of many offshoots, named for the song "Acid Tracks" by Phuture, a DJ collective.

Group Tripping in the Twenty-First Century

How we trip together keeps evolving, too. A kind of group trip has emerged that is like a stadium session, with fifty or more strangers sharing a large space and tripping together, all lying in rows on their mats with their blindfolds on. A young friend described it to me. After orientation and a lecture by the lead guide about giving in to the experience, participants settled onto their mats for an eight-hour adventure. A sound bath with live components controlled the trip, and the guide and his team kept an eye on everybody. The event cost from $500 to $1,000 per person, based on ability to pay. And it wasn't just psilocybin that people took but a double hippie flip: MDMA, then psilocybin, then MDMA, then psilocybin again. "It was a long night," my friend said. I thought it sounded risky, because, well, fifty people. But she assured me that everyone was well behaved. "Being with that many humans who were exploring their own consciousness is intense," she told me. "The shaman instills all with confidence they can do this work on their own," she said, "but the fact that everyone consents to doing this together is a powerful part of the experience."

Mushrooms and Extrasensory Perception

All those people tripping together can have a rather mysterious corollary. "Many times, I have had telepathic experiences on mushrooms," said Howard Sprouse. "It's one of the hallmarks of the psychedelic experience." Sprouse, who has a company called the Remediators, which uses plants and microbes—including fungi—to clean up polluted sites, is a member of a group with a dozen or so people, male and female, who have been tripping together for thirty years. They often share visions while tripping, he said, and sometimes even when they are not. Trippers have reported the phenomenon of telepathic group mind, calling it *grokking*, after the 1961 science fiction novel *Stranger in a Strange Land* by Robert A. Heinlein. It means to empathize so deeply with another, you merge with them.

When telepathy occurs between tripping people in a group, it makes some folks feel like their boundaries have been breached (telepathically) or they suddenly—sometimes uncomfortably—realize intimate things about their companions, like whether they are hiding an illness. Telepathic communication is a weird phenomenon where, after some confusion about whether they were speaking aloud or not, trippers might conclude they are conversing mind to mind, or that they have shared hallucinations. A fellow from Chicago told me a story about tripping in the parking lot of the Grand Canyon. He stayed in the car while his friends took in the view "because I couldn't discern where the solid ground stopped and the canyon began." In the rearview mirror he saw a police car had pulled up behind him. "At that moment I started inflating my head like a balloon until it filled the car, with my ears and face pressed against the windows. The police immediately drove away." Later, when he was reunited with his friends and they'd all come down, he asked the others if they'd seen the cops. One of them replied, "Yeah, but they drove off when you filled the car up with your head."

I've heard many versions of this anonymous post on Reddit: "I had a friend who said when he was on mushrooms, he saw a poster of Einstein on the wall wink at him. Just as he was going to tell a friend who was with him, his friend asked, 'Did you see Einstein wink?'" A young woman I know, Izzy, told me she witnessed two women tripping, and "at one point, their trips intertwined. They were on the same rocket ship. Watching them was really funny because you could see the moment their trips combined, and then all of a sudden, they were holding each other and pointing at random things together."

An early test of ESP while tripping was held at a Grateful Dead concert in 1971 in Port Chester, New York, where the researchers believed they had a large tripping population to work with (good guess). A big screen invited the audience to send telepathic messages to two "psychic sensitive people asleep in Brooklyn." The results didn't deliver. Science is a long way from finding a shared neurochemical action between the psychedelic experience and ESP. But a few researchers are pursuing it. David Luke is a psychologist at the University of Greenwich in the UK who studies the psychology of exceptional experiences. In a 2008 survey of 139 trippers, he found that 50 percent reported experiencing telepathy, and some even regularly practiced telepathic phenomena with a co-tripper. Another researcher, Petter Grahl Johnstad, had similar findings in 2020. He found that telepathy manifested in different ways: as an information exchange, where people communicated in images as well as in words; as an exchange of feeling states, called *telempathy* (a combination of *telepathy* and *empathy*); and as a kind of "telepathic unity where one could not differentiate one's own thoughts and feelings from those of the friend or partner."

But I wonder if what may seem like group ESP isn't sometimes just extreme suggestibility. My husband told me a story about how, as a teenager, he had tripped with a few friends at a ski resort and they were driving home (the driver was sober) when the passengers convinced themselves they'd just had a car accident. They pulled over and inspected the car and were surprised to discover that, despite their collective recollection, nothing had happened at all.

Finding Your Community

I met Aaron Paul Orsini of the Autistic Psychedelic Community in person at the Horizons: Perspectives on Psychedelics conference in 2022, and he crashed in my little guest room for a few nights. One of those nights, he suggested he invite a few people from the psychedelic scene over for a drink, which sounded great to me. I ended up cooking pasta for about a half dozen women, all of them graceful and refreshingly forthright. That's where I met Pamela Jackson, founder of the Psychedelic Sisterhood, who told me she was raised Christian evangelist in Indiana but became atheist because "the Christianity I grew up with said if you don't believe in this very specific way, you weren't Christian." Through psychedelics, she "received the spiritual relationship I was supposed to feel in church," she said. I met the writer Rachel Nuwer, author of a book on MDMA called *I Feel Love*, who was feeling a little trepidation about an ibogaine contact she was meeting in Gabon and so let me make her a martini. I also met Kat Lakey, lovely and reticent as a dove, who admired a mycological drawing on the wall and said it was the perfect vibe for the Athenaeum, her soon-to-open psychedelic library, coworking, and event space.

The Athenaeum idea grew out of the Psychedelic Assembly, a conference organized by Lakey and Orsini in New York City in 2022. It attracted prominent folks in the psychedelic scene, like (former) LSD chemist William Leonard Pickard and Dr. Julie Holland, author of *Good Chemistry*, one of the psychedelic genre's most popular books. The conference's success inspired Lakey to establish a permanent home for the city's psychedelic community. Today, the Athenaeum is housed in the same space where the conference was held, on the ground floor of a blue-painted tenement in Midtown Manhattan.* (That's a rather incongruous address for a psychedelic rec room: it's next door to Sparks Steak House, famous for its prime sirloin and being the killing ground of wiseguy Paul Castellano in 1985.)

Since Lakey and her partners, Abby Lyall and Suzy Baker, opened the Athenaeum in 2023, it has acquired a substantial library of 1,500 books on psychedelics and related titles; established a bookstore and café; and continued to offer constant programming of all sorts, from book signings and panel discussions on psychedelic topics to comedy acts and Find the Others First Fridays, a monthly social gathering. "When I finally found this community," said Lyall, who has since left the project, "it became obvious to me how critically important the community aspect was."

For me, the in-person aspect of the Athenaeum is a welcome antidote to the barrage of Zoom seminars and conferences and summits on psychedelic subjects and the relentless marketing of opinions and trainings and products, all of them vying for my wallet's attention. It's nice to be back in a library. Indeed, I wrote part of this chapter in the Athenaeum, in a room full of plants and books, natural history objects, and mismatched furniture, like a cross between a hippie café and an ethnobotanist's

* The space is called ClearLight—a reference, said the owner of the building to the *New York Times*, to a "major LSD distribution network in California back in the day."

office. At one point while typing, Gina Mostafa, a gracious Egyptian woman who works in health equity and volunteers for the Athenaeum, asked if I had any music preferences. I said no.

"Okay," she said. "I'll play metal."

I attended the Athenaeum's opening-night celebration, and the place was packed. A variety of programming was planned: performances, readings (Dennis McKenna read from the reissue of his book *The Brotherhood of the Screaming Abyss*), presentations—some body painting—all of it happening at once in a kind of chaotic flow. And all kinds of folks were mixed together: hippies and fashionable New Yorkers, investment bankers and baristas, people dressed like priests and priestesses, gray-haired baby boomers who brought their years of experience in the underground psychedelic scene, and young people who brought energy and new ideas. And it occurred to me that the psychedelic community is like that: wildly diverse, if at times siloed by their circumstances and needs, but ultimately a community anyone can fit into.

It's a Big Tent

The community of psychonauts today is a big tent. There are the microdosers and museum dosers, the therapeutic users, the meadow trippers, and the deep divers into ego dissolution. In the tent are those who trip solo in the woods or in their bedrooms, or with others on Zoom; those who are members of psychedelic churches or religious integration groups, or who create very personal ceremonies with close friends; those who trip in massive groups, the psychedelic version of collective weddings or graduations. The tent includes the ravers and Burners and other festivalgoers, and the folks who take care of them if their trip gets dicey. Indeed, the sitters and facilitators, therapists and researchers who support trippers are in the tent as well, whether they trip or not. So are the advocates, the lawyers, the hunters, and the cultivators. Musicians and those who make instruments, artists, filmmakers, philosophers, and writers all contribute to the psychedelic community today. Even though white people dominate the fields of psilocybin industry and science, the bulk of the felt psychedelic experience includes myriad races, genders, and faiths. And in a place of honor, the indigenous people and their ancestors, whose generations of work with magic mushrooms represent the only truly long-term study anyone has about these substances.

I spoke with many of these people about their psilocybin experiences, read their (often madcap) trip reports, and took the mushrooms myself in order to expand my understanding of how we use magic mushrooms today. I mined the internet forums, a rich and wacky vein of insights whose members ranged from thoughtful and supportive to, in my view, utterly reckless. I also attended conferences and summits and webinars, read books and articles and blog posts, and had to ask myself every step of the way, *Is this educational or just marketing hoo-ha?*

Over the course of my research, I learned there are many shared phenomena—they are what's fueling the science—but I also realized there are no absolutes when it comes to magic mushrooms. Every trip seems to exist on a spectrum of possibilities. A trip can provide wondrous insights and nightmarish visions, great empathy and deep sorrow, healing and harm, and even nothing at all. The most consistent takeaway for me, and what I've tried to acknowledge over and over, is the utterly subjective nature of the psychedelic experience.

Trippers are curious and hopeful: curious to learn about the workings of their brains and hopeful they will be relieved of distress, see God, have a really good time. But curiosity and hope are where our commonality ends. We are all different: our individual histories, coping mechanisms, and chemistries make for psychedelic experiences as varied as the people who try them. That's why I cannot say what will happen to you, or anyone else, if you eat a magic mushroom.

The only thing I can confidently do is share information. What you do with it is up to you. But I think magic mushrooms should be considered judiciously, even if your intention is to spend a very unserious afternoon sitting in a daisy-filled meadow. Because at the end of the day, what constitutes a good trip is all about what happens in *your* mind.

Epilogue

Over the course of about a year, I took magic mushrooms more times than I had in all my years previous, and I can honestly say that, overall, it has been a very good trip indeed. I believe the large doses of mushrooms I took helped me identify my tendency to be dogmatic and judgmental. The mushrooms helped unpeel me, like accelerated therapy. Microdosing helped me get past—and stay past—those tendencies.

Here's an example of where I was and where, post-trips, I ended up. We have a cabin in Colorado, and every couple of years, we throw a pig roast there. This involves stoking a tremendous pit fire down to ashes, preparing the pig, wrapping it up in cheesecloth, and burying it among the hot coals for about twelve hours. It is a tremendous hassle, but when it's done, the meat is tender and succulent, and we spend the afternoon sitting around picnic tables drinking from a keg of beer and eating pork tacos with our neighbors. My husband Kevin's family, including his ninetysomething mother, drive up from Santa Fe for the festivities. A few years ago, our son had offered to come out and manage the pig roast, along with a few friends. I was all for passing the torch, but I ended up keeping a tight rein on, well, everything. I gave Mo and his friends very detailed directions on how to prep the fire and the pig, to the point where they'd just come around for more orders rather than thinking through the next steps themselves. I think controlling the narrative so tightly squelched their initiative.

As a result, I kind of dreaded the pig roast because it was so much work. Or rather, I made it so much work for me. I also cooked a few dinners for my husband's family, who love to eat, and I'd serve meals the way I was raised by my dogmatic Italian relatives: a pasta dish, followed by a meat or vegetables, then a salad. Three plates, in that order. My western family would wait for the courses, somewhat resignedly, the way you wait for a driver to get himself settled in and buckled up before he pulls out of a parking space you want. On the occasion when I would have to put the pasta, meat, and salad on the table at the same time, they inevitably piled everything into their pasta bowl. I served those meals a certain way, and for decades, I was irritated that they never got with the program.

The year of my trips, however, everything changed. Our son offered to manage the pig roast again, and I gave him the farmer's contact information and then forgot about it. Twenty-six of his friends showed up at the cabin, in cars and trucks, on motorcycles and by plane, women, men, dogs, and they set up a little tent city near our dilapidated but picturesque barn. With Kevin's family, we had about thirty people sleeping at the house; that's a lot of flushes for our two-bath cabin, so rather than developing bathroom use rules, I rented a portable toilet and had it installed in the barn. After that, I sat back and let the pig roasters ask me what they needed. Okay to make pancakes? I showed them where the flour and baking powder were. What to stuff in the pig? I said, "Look around the garden and orchard, and see what is growing," and so they filled the carcass with branches of oregano and sweet yellow plums. They asked about side dishes, and I gave them a couple of recipes and then spent the day antiquing with my sister-in-law. I made dinner for the old folks (the younger generation had an ongoing barbecue scene on one of the cabin porches) and served pasta with bacon, beef stew with porcini, and avocados with red onions, enjoying a glass of wine while everyone ate in whatever order they liked. The young folk stoked a big fire in the pit for most of the day, and I didn't worry about them setting the mesa on fire. By midnight, it had burned down to coals, and they buried the pig and patted down the earth with the backs of their shovels, and I slept through it all.

In the morning, I went out to check on the pit. In the past, I'd spent mornings picking up beer bottles and cigarette butts, but not this time. All around the pit, it was tidy, just a few buckets of strategically placed water. When I walked back inside, I saw someone emerge from the bathroom. Despite a pre-pig-roast meeting where the young folks were urged to use the portable toilet, Mo's friends really preferred a proper restroom. That's when I noticed the water rising in the toilet and slipping over the lip of the bowl. Once alerted, Mo and his friends got to work on it, but as they plunged, brown water started to burble up from the drain in the shower. They mopped that up, too, and we put red tape across the door, and then someone washed a coffee mug in the kitchen sink and the shower filled with brown gunk again. By midmorning, we had one working toilet, no draining kitchen sink or dishwasher, no clothes washer, and fifty people expected later in the afternoon.

Our guests managed: the elderly pig roasters used our one functioning bathroom (on a separate septic). Everyone else made do with the portable toilet, which we had strewn with branches of wild sage. But here's the thing: I never got upset. I guarantee in the past, I would have been a ball of nerves, on the edge of snapping, unable to laugh it off. But instead, the whole situation was just kind of ridiculous, made even sillier when the company that showed up to drain our septic the next day consisted of two old cowboys whose business was called Cheerful Cesspools. The main thing I remember was the pig tasted fantastic.

After everyone left—Mo's crew scattered across the country and the family headed back to New Mexico—Kevin told me his brother had said he'd never seen me so chill and relaxed.

"It's the mushrooms," Kevin replied.

This is what I think happened to me: The large trips helped me recognize deep-seated uncertainties, like fear of inherited illness, and put in perspective what is worth getting upset about and what is not. This led me to feeling less attached to convictions that were more about style than substance, like at what point in a meal you eat your salad. And once I stopped getting upset about salad, I started to feel more confident about the things I did care about, like letting the next generation run the show.

Letting go of superficial convictions meant I didn't have to be right to be okay. I found I wasn't as reliant on praise nor as upset by criticism. I became less occupied with defending or promoting one position or another, and that created an opportunity in me to be more sensitive to other people, to be a better listener, to be curious. As a result, my interactions seem easier and more authentic. Mary Cosimano, who has worked in Johns Hopkins's psychedelic program for years, said in a TED Talk that she had witnessed hundreds of tripping patients, and at the core of their experiences was "a reconnection with their authentic selves."

I credit microdosing with helping me learn what it feels like when I am my authentic self, dialed in to the moment, compassionate and attentive. Now when I start to overthink or sputter bullshit—behaviors I believe are rooted in anxiety—I can *feel* it coming. I recognize a hot sensation in my chest that precedes these obnoxious behaviors, and once warned, I am better able to stop and self-correct—not always, but better. My friend the therapist Meg whom I met at Blue Portal calls this "the quick turnaround." The great surprise, however, was it has also lifted much of my trepidation about getting older, perhaps because I feel more comfortable just being me.

By the time you are reading this book, I will be a senior citizen. Most of the science on elderly people and psilocybin investigates the extremes: dementia, depression, end of life. The only place to look for information on how magic mushrooms might benefit older people who are just dealing with the arc of life is by asking folks who have used these drugs, like the poet laureate of the Telluride Mushroom Festival, Art Goodtimes.

Art is like an ancient elf, capering and merry; the festival's cheerful shaman. In the fall of 2023, he presided over the wedding of two mycologists—the bride carried fungi—and in his remarks, which were like a prayer poem, he blessed the newlyweds and their place in the great cycle of life, death, rot, and rebirth. After the festivities, I asked him if he had any thoughts on magic mushrooms and aging. For Goodtimes, psychedelics (which he refers to as *entheogens*) have broken down his preconceptions about aging and death. "I started as a very religious young man, and entheogens helped me escape that straitjacket. Ever since, I have walked a different path in life than most of my colleagues. I believe this is because I was able to find my own true self after my entheogenic experiences. This has carried over into my life as I've aged. My recent challenges with cancer and taking hospice care of my father and my wife, and being there when they passed, singing them into the mystery, have led me to accept death—as mysterious as it is. And that acceptance has led me to enjoy aging, to enjoy being an elder, to enjoy the full cycle of life."

Senior citizens aren't as active on internet forums, but occasionally, someone points out the special benefit they see psilocybin occasioning. "The problem for seniors is more than hardening of the arteries, it is the hardening of attitudes," wrote one person on Shroomery. "As we age, we become more cautious, more conservative, more scared. Mushrooms can be the best way to change this for seniors."

If you are old and it doesn't feel good to be old, the benefit of taking the mushrooms may be to help shake off entrenched behaviors. Thought patterns that have accumulated over the years, that bog us down, may disassemble and lose their agency. And if you are young, the mushrooms may help ensure you avoid those rigid ideas in the first place. But whatever your age, the magic in magic mushrooms is their potential to put you in tune with your true self.

Many years ago, in my book about mushroom hunting, *Mycophilia*, I wrote, "Mushrooms were the window by which I came to understand nature in a deeper way." I'm going to recast that sentiment: for me, magic mushrooms were the window by which I came to understand *myself* in a deeper way.

Acknowledgments

I've paid attention to the magic mushroom scene for a long time. I wrote about it in *Mycophilia* and teach a class on psychedelic mushrooms at the New York Botanical Garden. But my timing for writing a book didn't come together until Pam Krauss, an extraordinary editor, told me Flatiron was interested in magic mushrooms. So first, huge thanks to Pam. Seasoned editors like her make books better.

I also want to thank the exceptional team at Flatiron: Bob Miller, Megan Lynch, Malati Chavali, Will Schwalbe, Julie Will, Sam Zukergood, Kate Lucas, Morgan Mitchell, Jason Reigal, Chris Smith, Maris Tasaka, Keith Hayes, and Laury Frieber. Thanks to my agent, Angela Miller, and my friend Kris Dahl for their great advice. Many thanks to the folks who reviewed early drafts: Jordan Jacobs, Rev. Naomi Schulz, and Aaron Paul Orsini; and particular sections: Paul Sadowski and Alan Rockefeller. I showed everyone I interviewed their quotes in context, and most people responded with corrections and kind words. Thank you to the many scientists and advocates who helped me understand your work, especially those who gave me so much of your precious time. (I'm talking about you, Peter Hendricks.) That said, any errors in the book are mine.

An earlier draft of this book included the other psychedelic mushrooms, mostly the fly agaric (various *Amanita* species). I'm sorry we couldn't include that material, but many thanks to those folks who helped me with it, especially Irene Liberman, Mariana Villani, Amanita Dreamer, Kevin Feeney, Spike

Mikulski, John Michelotti, Brandon Pitcher, Eric Alexander Yanasak, and the folks at Amanita Muscaria Science and Magic.

This book is mainly built on anecdotal reporting—the experiences of people who use magic mushrooms. Internet forums, both focused, like Clusterbusters, and more freewheeling, like Shroomery and Reddit, offer a wealth of trip reports. Okay, with some of these forums, you must wade through lots of silly memes and some colorful commentary, but it is so worth it. Many thanks to those of you who have felt compelled to share your stories. You are helping everyone. I especially want to thank those named and unnamed trippers who shared their stories with me personally. This book is laced with your candidness and generosity.

I also owe a debt of gratitude to the many friends and colleagues who are not named in this book but whose advice and insights directed my research and understanding: Art Chandler, Jonathan Rosenthal, Randy Polumbo, Alexandra Adams of Stanford University, Danni Peterson of the Association of Entheogenic Practitioners, Carla Asquith, Farid Alsabeh, LLP, Sebastian Carosi, Britt Bunyard, David Shearer, Elizabeth Johnson, Peter Schubert, Leslie Bryan, Hadas Alterman of the American Psychedelic Practitioners Association, Dan Parnooch, Steve Rubick, Daniel Winkler, Sandra Dreisbach, Jane Cyphers, Thomas Hughes, Brendan Doyle, Larry Kirkland, Abbi Klein, LCSW, Jeff Bernstein, Lee Woodruff, Susan Viebrock, Chi at Mushroom Tao, Elisabeth Ottolini, Wesley Fordyce, Carson Bone, Mo Bone, Alan Eisner, the Mirage Garage family, the folks at Horizons 2022 and Psychedelic Science 2023, and Zendo. Special thanks to Kevin Bone, who not only shared his tales but, to my endless gratitude, shares my life.

Finally, I'd like to acknowledge the work of my fellow writers in this field. Thank you, Jules Evans of Ecstatic Integration, and the staff writers at *DoubleBlind*, *Vice* (especially reporting by Shayla Love), Psychedelics Today, the Vault at Erowid, Psychedelic Spotlight, Psychedelic Science Review (Barbara Bauer, you are wonderful), Psychedelic Alpha, the Microdose, Lucid News (love the psychedelic auntie), Hyphae News, *Psychedelic Medicine* (the journal), and the merry genius of Adam Aronovich of @Healingfromhealing.

Notes

Author's Note

xvii *medicine implies sickness*: "Kilindi Iyi—High-Dose Mushrooms Beyond the Threshold," Breaking Convention, May 21, 2019, YouTube video, 42:06, https://www.youtube.com/watch?app=desktop&v=ejdKeghBhNs.

xvii *"and we are all seeking"*: *Bad Faith with Briahna Joy Gray* podcast, episode 195, "How to Do Drugs (w/ Dr. Carl Hart)," July 21, 2022, https://badfaith.libsyn.com/size/5/?search=Carl+Hart.

1: Considering Psilocybin

5 *"good-grade pill"*: Alan Schwartz, "Risky Rise of the Good-Grade Pill," *New York Times*, June 9, 2012.

7 *"Go on to any psychedelic reddit page"*: Ed Prideaux, "On the Qualitative Richness of Psychedelic Reddit," Ecstatic Integration, October 13, 2023, https://www.ecstaticintegration.org/p/on-the-qualitative-richness-of-psychedelic.

8 *"It is our greater or lesser"*: Elizabeth Redfern, *The Music of the Spheres* (New York: Jove, 2001), 324.

9 *no one really knows*: Torstein Passie et al., "The Pharmacology of Psilocybin," *Addiction Biology* 7, no. 4 (2002): 357–64.

9 *Acids break apart*: For a good video on how toxins (including psilocybin) move from gut to liver to bloodstream, see "Liver Function 1, Filter Between Gut and Blood," Dr. John Campbell, June 9, 2019, YouTube video, 7:15, https://www.youtube.com/watch?v=1wvVwBRjYMU.

10 *"a new, alien"*: Andy Letcher, *Shroom: A Cultural History of the Magic Mushroom* (London: Faber and Faber, 2006), 17.

12 *it just helps you feel not shitty*: Ayelet Waldman, *A Really Good Day: How Microdosing Made a Mega Difference in My Mood, My Marriage, and My Life* (New York: Anchor, 2018), x.

12 psycholytic therapy *was studied*: Torsten Passie, Jeffrey Guss, and Rainer Krähenmann, Lower-Dose Psycholytic Therapy A Neglected Approach," *Frontiers in Psychiatry* 13 (December 2022).

14 *"unfolding process"*: Ingmar Gorman et al., "Psychedelic Harm Reduction and Integration: A Transtheoretical Model for Clinical Practice," *Frontiers in Psychology* 12 (March 2021).

15 *"tool for exploration"*: "Kilindi Iyi Tribute Video: Psychedelic Pioneer with High-Dose Psilocybin," Chacruna Institute, July 16, 2020, YouTube video, 6:46, https://www.youtube.com/watch?v=LoDZ-zq_nnM.

17 *as you age, psychedelic trips*: Kwonmok Ko et al., "Predicting the Intensity of Psychedelic-Induced Mystical and Challenging Experience in a Healthy Population: An Exploratory Post-Hoc Analysis," *Neuropsychiatric Disease and Treatment* 19 (2023): 2105–13.

17 *"may be the single most important factor"*: Andrew Weil, "The Use of Psychoactive Mushrooms in the Pacific Northwest: An Ethnopharmacologic Report," *Botanical Museum Leaflets, Harvard University* 25, no. 5 (1977): 131–49.

20 *potential to induce valvular heart disease*: Jane C. Hu, "Heart Risks, Serotonin, and Fen-Phen: 5 Questions for 5-HT Receptor Researcher Bryan Roth," *Microdose*, September 18, 2023, https://themicrodose.substack.com/p/heart-problems-serotonin-and-fen.

2: Brains on Mushrooms

26 *It's one of the factors*: "Brain Basics: The Life and Death of a Neuron," National Institute of Neurological Disorders and Stroke, https://www.ninds.nih.gov/health-information/public-education/brain-basics/brain-basics-life-and-death-neuron.

26 *is thought to interact primarily*: Robin Carhart-Harris and David Nutt, "Serotonin and Brain Function: A Tale of Two Receptors," *Journal of Psychopharmacology* 31, no. 9 (2017): 1091–120.

27 *This receptor's signaled pathway*: Roberta Strumila et al., "Psilocybin, a Naturally Occurring Indoleamine Compound, Could Be Useful to Prevent Suicidal Behaviors," *Pharmaceuticals* 14, no. 12 (2021): 12–13.

27 *Psilocin has a low affinity*: Thomas S. Ray, "Psychedelics and the Human Receptorome," *PLOS ONE* 5, no. 2 (2010): e9019.

27 *5-HT2A serotonin receptor is the most studied*: F. Tylš, T. Páleníček, and J. Horáček, "Psilocybin—Summary of Knowledge and New Perspectives," *European Neuropsychopharmacology* 24, no. 3 (2014): 342–56.

27 *norpsilocin does indeed interact with 5-HT2A*: "Baeocystin," Psychedelic Science Review, https://psychedelicreview.com/compound/baeocystin.

27 *psilocin doesn't act only on serotonin 5-HT2A*: Maxemiliano V. Vargas et al., "Psychedelics Promote Neuroplasticity Through the Activation of Intracellular 5-HT2A Receptors," *Science* 379, no. 6633 (2023): 700–6.

28 *regions of the brain—and there are many*: B. T. Thomas Yeo et al., "The Organization of the Human Cerebral Cortex Estimated by Intrinsic Functional Connectivity," *Journal of Neurophysiology* 106, no. 3 (2011): 1125–65.

29 *One consistently implicated area*: James J. Gattuso et al., "Default Mode Network Modulation by Psychedelics: A Systematic Review," *International Journal of Neuropsychopharmacology* 26, no. 3 (2023): 155–88.

29 *"how we relate"*: Michael Pollan, *How to Change Your Mind: What the New Science of Psychedelics Teaches Us About Consciousness, Dying, Addiction, Depression, and Transcendence* (New York: Penguin, 2018), 391.

29 *fMRI measurements of the DMN*: R. Smausz, J. Neill, and J. Gigg, "Neural Mechanisms Underlying Psilocybin's Therapeutic Potential—The Need for Preclinical in Vivo Electrophysiology," *Journal of Psychopharmacology* 36, no. 7 (2022): 781–93.

30 *Some folks have reported a reduction*: Robin Carhart-Harris et al., "Psilocybin for Treatment-Resistant Depression: fMRI-Measured Brain Mechanisms," *Scientific Reports* 7 (2017): 13187; R. E. Daws et al., "Increased Global Integration in the Brain After Psilocybin Therapy for Depression," *Nature Medicine* 28 (2022): 844–51.

30 *Robin Carhart-Harris has described*: Nicola Davidson, "The Struggle to Turn Psychedelics into Life-Changing Treatments," *Wired (UK)*, December 5, 2018, https://www.wired.co.uk/article/psychedelics-lsd-depression-anxiety-addiction.

31 *tripper's consciousness consists of*: Huston Smith, *Cleansing the Doors of Perception: The Religious Significance of Entheogenic Plants and Chemicals* (New York: Penguin Putnam, 2000), 92.

32 *the higher the dose*: Roland R. Griffiths et al., "Psilocybin Produces Substantial and Sustained Decreases in Depression and Anxiety in Patients with Life-Threatening Cancer: A Randomized Double-Blind Trial," *Journal of Psychopharmacology* 30, no. 12 (2016): 1181–97.

33 *a hallucination is defined*: Oliver Sacks, *Hallucinations* (New York: Knopf, 2012), 7.

33 *You may see the world through*: P. C. Bressloff et al., "Geometric Visual Hallucinations, Euclidean Symmetry and the Functional Architecture of Striate Cortex," *Philosophical Transactions of the Royal Society London B* 356, no. 1407 (2001): 299–330.

34 *"entities"*: James Fadiman, *The Psychedelic Explorer's Guide: Safe, Therapeutic, and Sacred Journeys* (Rochester, VT: Park Street Press, 2011).

34 *María Sabina called the "holy children"*: K. Williams et al., "Indigenous Philosophies and the 'Psychedelic Renaissance,'" *Anthropology of Consciousness* 33 (2022): 506–27.

35 *The phenomenon is not really understood*: "What Is Synesthesia?," *Scientific American*, September 11, 2006, https://www.scientificamerican.com/article/what-is-synesthesia.

35 *Some scientists have proposed*: Ethan Bilby, "Do We All Have Synaesthesia?," *Horizon*, October 20, 2015.

37 *our brains don't spend capital*: Scott McLemee, "Wrinkles in Time," *Inside Higher Ed*, August 30, 2018, https://www.insidehighered.com/views/2018/08/31/review-marc-wittmann-altered-states-consciousness-experiences-out-time-and-self.

37 *adults, regardless of culture*: McLemee, "Wrinkles in Time."

37 *"a reduced autobiographical memory load"*: Marc Wittmann and Nathalie Mella, "Having Children Speeds Up the Subjective Passage of Lifetime in Parents," *Timing & Time Perception* 9, no. 3 (2021): 275–83.

37 *"When you are a child"*: David Eagleman, "Brain Time," Edge, June 23, 2009, https://www.edge.org/conversation/david_m_eagleman-brain-time.

37 *genes that regulate*: O. V. Sysoeva, A. G. Tonevitsky, and J. Wackermann, "Genetic Determinants of Time Perception Mediated by the Serotonergic System," *PLOS ONE* 5, no. 9 (2010): e12650.

38 *"The predominant feeling"*: Fadiman, *The Psychedelic Explorer's Guide*, 17.

42 *"Ego-Dissolution and Psychedelics"*: Matthew M. Nour et al., "Ego-Dissolution and Psychedelics: Validation of the Ego-Dissolution Inventory," *Frontiers in Human Neuroscience* 10 (2016).

42 *He described the ego as*: "fMRI Brain Networks in 10 Minutes | Default-Mode Network and Others Explained," Psychedelic Scientist, September 2, 2020, YouTube video, 9:23, https://www.youtube.com/watch?v=7Cto2yr5NQY.

43 *"We are left"*: "fMRI Brain Networks," YouTube.

43 *some may feel worn-out*: Fadiman, *The Psychedelic Explorer's Guide*, 223.

45 *"philosophical entertainer"*: "Alan Watts," Wikipedia, https://en.wikipedia.org/wiki/Alan_Watts.

45 *"if you get the message"*: Quote by Alan Watts, Goodreads, https://www.goodreads.com/quotes/6639601-if-you-get-the-message-hang-up-the-phone-for.

45 *Researchers have associated*: Griffiths et al., "Psilocybin Produces Substantial and Sustained Decreases."

45 *The prevailing hypothesis*: Matthew M. Nour and Robin Carhart-Harris, "Psychedelics and the Science of Self-Experience," *British Journal of Psychiatry* 210 (2017): 177–79.

46 *The paper "Psilocybin Produces Substantial"*: Roland R. Griffiths et al., "Psilocybin-Occasioned Mystical-Type Experience in Combination with Meditation and Other Spiritual Practices Produces Enduring Positive Changes in Psychological Functioning and in Trait Measures of Prosocial Attitudes and Behaviors," *Journal of Psychopharmacology* 32, no. 1 (2018): 49–69.

47 *"strengthen, loosen, and reorganize"*: Clare Watson, "The Psychedelic Remedy for Chronic Pain," *Nature*, September 28, 2022, https://www.nature.com/articles/d41586-022-02878-3.

47 *an acute "hyperplastic" state*: Kate Godfrey, "Electrophysical EEG Markers of Human Neuroplasticity in Psychedelic Research," Interdisciplinary Conference on Psychedelic Research 2022, https://community.open-foundation.org/c/icpr-2022-recordings/.

3: Bad Trips

51 *characterized by intense frustration*: Rachael Petersen, "A Theological Reckoning with 'Bad Trips,'" *Harvard Divinity Bulletin*, Autumn/Winter 2022.

52 *That's in line with data collected*: Jules Evans et al., "Extended Difficulties Following the Use of Psychedelic Drugs: A Mixed Methods Study," SSRN, June 21, 2023, https://ssrn.com/abstract=4505228.

53 *emotionally distressed off-duty pilot*: Michael Levenson, "Pilot Who Disrupted Flight Said He Had Taken Psychedelic Mushrooms, Complaint Says," *New York Times*, October 24, 2023.

53 *Richards told a story*: "Bill Richards on Psychedelics Sacred Knowledge at CIIS," Reese Jones, July 6, 2016, YouTube video, 52:23, https://www.youtube.com/watch?v=EsgKUglCI7g.

54 *"What is so important here"*: William A. Richards, *Sacred Knowledge: Psychedelics and Religious Experiences* (New York: Columbia University Press, 2016), 86.

55 *"non-specific amplifiers"*: David Fuller, "Our Shared Pandemic Psychedelic Trip, with Erik Davis," Medium, July 15, 2020, https://medium.com/rebel-wisdom/our-shared-pandemic-psychedelic-trip-with-erik-davis-10aa1e8ac366.

56 *There is simply too much*: Shayla Love, "At-Home Genetic Testing Can't Tell You If You're Going to Have a Bad Trip," *Vice*, August 20, 2021, https://www.vice.com/en/article/y3d4xb/genetic-testing-predict-bad-psychedelic-trip-halugen.

56 *psychedelics rarely trigger*: A. K. Schlag et al., "Adverse Effects of Psychedelics: From Anecdotes and Misinformation to Systematic Science," *Journal of Psychopharmacology* 36, no. 3 (2022): 258–72, https://doi.org/10.1177/02698811211069100; "David Nichols: How Do Psychedelics Work?," OPEN Foundation, August 31, 2022, YouTube video, 15:43, https://www.youtube.com/watch?v=UCDS9n351dw.

56 *Emerging research suggests*: "COMP360 Psilocybin Therapy Shows Potential in Open-Label Study in Type II Bipolar Disorder Presented at ACNP," GlobeNewswire, December 8, 2022, https://www.globenewswire.com/news-release/2022/12/08/2569941/0/en/COMP360-psilocybin-therapy-shows-potential-in-open-label-study-in-type-II-bipolar-disorder-presented-at-ACNP.html.

59 *2020 survey conducted by Dr. Symon Beck*: "EGA Microdose Webcast 6: Woodlover Paralysis with Symon Beck and Caine Barlow," Entheogenesis

Australis - Entheo TV, September 1, 2021, YouTube video, 1:35:50, https://www.youtube.com/watch?v=JQNzJvDEKIs.

60 *one hypothesis credits the effects*: Barbara Bauer, "Wood Lover Paralysis from Magic Mushrooms: The Aeruginascin Hypothesis," Psychedelic Science Review, April 1, 2019, https://psychedelicreview.com/wood-lover-paralysis-from-magic-mushrooms-the-aeruginascin-hypothesis.

61 *Denver, which has legalized psilocybin*: Sara Gail and Bryan Lang, "MAPS Harm Reduction Department Initiates Psychedelic Response Training for Denver First Responders," MAPS *Bulletin* 31, no. 2 (2021).

61 *they happen in clinical trials, too*: Grace Browne, "The Harms of Psychedelics Need to Be Put into Context," *Wired*, November 3, 2022, https://www.wired.com/story/psychedelics-side-effects.

61 *they just don't apply to every bad trip*: J. J. Breeksema et al., "Adverse Events in Clinical Treatments with Serotonergic Psychedelics and MDMA: A Mixed-Methods Systematic Review," *Journal of Psychopharmacology* 36, no. 10 (2022): 1100–17.

62 *due to follow-up constraints*: Breeksema et al., "Adverse Events"; Roland R. Griffiths et al., "Psilocybin Produces Substantial and Sustained Decreases in Depression and Anxiety in Patients with Life-Threatening Cancer: A Randomized Double-Blind Trial," *Journal of Psychopharmacology* 30, no. 12 (2016): 1181–97.

62 *various experts in the field*: Jules Evans, "The Cost of Utopia," Ecstatic Integration, October 27, 2023, https://www.ecstaticintegration.org/p/the-cost-of-utopia.

62 *"serious effort to examine bad trips"*: Petersen, "A Theological Reckoning."

63 *a study in 2023 suggested the co-use*: R. J. Zeifman et al., "Co-use of MDMA with Psilocybin/LSD May Buffer Against Challenging Experiences and Enhance Positive Experiences," *Scientific Reports* 13, no. 1 (2023), article 13645.

64 *anecdotal consensus is*: G. Valeriani et al., "Olanzapine as the Ideal 'Trip Terminator'? Analysis of Online Reports Relating to Antipsychotics' Use and Misuse Following Occurrence of Novel Psychoactive Substance-Related Psychotic Symptoms," *Human Psychopharmacology* 30, no. 4 (2015): 249–54.

64 *It may be their trips*: Jules Evans, "Do Indigenous People Have Totally Different Trips to Westerners?," Medium, April 23, 2019, https://julesevans.medium.com/do-indigenous-people-have-totally-different-trips-to-westerners-de7ba3d8807f.

65 *"Donald Trump pills"*: The Loop (@WeAreTheLoopUK), "Donald Trump pills causing serious medical problems across the UK this

weekend," Twitter, August 24, 2019, 2:51 p.m., https://twitter.com /WeAreTheLoopUK/status/1165335945400455168.

65 *"somebody watches a Netflix show"*: "Discussing Harm Reduction in the Wake of Netflix's Docuseries How to Change Your Mind - Fireside," Psychedelic Spotlight, July 19, 2022, YouTube video, 29:05, https:// www.youtube.com/watch?v=LIv2HNOFxfo.

65 *perhaps even the personality disorders*: A. Marrocu et al., "Psychiatric Risks for Worsened Mental Health After Psychedelic Use," PsyArXiv, October 17, 2023, https://doi.org/10.31234/osf.io/2e34t.

4: Keeping Trips on Course

68 *The term* set and setting, *wrote*: Ido Hartogsohn, "Set and Setting in the Santo Daime," *Frontiers in Pharmacology* 12 (May 2021), https:// doi.org/10.3389/fphar.2021.651037.

69 *He and a few others*: R. Gordon Wasson, "Seeking the Magic Mushroom," *Life*, May 1957, https://bibliography.maps.org/bibliography /default/resource/15048.

70 *Wasson's article laid out*: Wasson, "Seeking the Magic Mushroom."

71 *Their antics*: Gary Lincoff and D. H. Mitchel, *Toxic and Hallucinogenic Mushroom Poisoning: A Handbook for Physicians and Mushroom Hunters* (New York: Van Nostrand Reinhold, 1977), 105.

71 *disrupted the village's social dynamics*: Paul Roberts, "Huautla de Jiménez: In the Footsteps of María Sabina and John Lennon," *livingandworkinginmexico* (blog), August 28, 2010, https://livingandworkinginmexico .wordpress.com/2010/08/28/huautla-de-jimenez-in-the-footsteps-of-maria -sabina.

71 *"stigma of being involved"*: Heriberto Yépez, "Re-Reading María Sabina," *Poems and Poetics* (blog), October 13, 2010, http://poemsandpoetics .blogspot.com/2010/10/heriberto-yepez-re-reading-maria-sabina.html.

71 *"She had to fulfill the work"*: Osiris García Cerqueda, "Magic Mushrooms, Memory and Resistance in the Sierra Mazateca," Chacruna Institute, August 18, 2020, https://chacruna.net/mushrooms-resistance -sierra-mazateca.

72 *"From an indigenous perspective"*: Konstantin Gerber et al., "Ethical Concerns About Psilocybin Intellectual Property," *ACS Pharmacology & Translational Science* 4, no. 2 (2021): 573–77.

72 *"If a plant has been used"*: Terence McKenna, *Food of the Gods: The Search for the Original Tree of Knowledge—A Radical History of Plants, Drugs, and Human Evolution* (New York: Bantam, 1992), 247.

73 *"accountability for perpetuation"*: Yuria Celidwen et al., "Ethical Principles of Traditional Indigenous Medicine to Guide Western Psychedelic Research and Practice," *Lancet* 18 (2023), article 10044109.

76 *Serotonin syndrome's symptoms*: For a good general overview of serotonin syndrome, see Georgina Tiarks, "Serotonin Syndrome: What It Is, Causes, Signs, Symptoms, and More," Osmosis, https://www.osmosis.org/answers/serotonin-syndrome.

76 *Serotonergic psychotropics like*: B. Malcolm and K. Thomas, "Serotonin Toxicity of Serotonergic Psychedelics," *Psychopharmacology (Berlin)* 239, no. 6 (2022): 1881–91.

76 *SSRIs may interfere*: Camile Bahi, "Antidepressants and Psychedelics," Mind Foundation, September 11, 2020, https://mind-foundation.org/psychedelic-antidepressant-interactions.

76 *(MAOIs) can cause severe serotonin syndrome*: Malcolm and Thomas, "Serotonin Toxicity." There aren't many places where you can find summaries of interactions. Here's one: "Can You Take Shrooms on Antidepressants?," Synthesis, https://www.synthesisretreat.com/psilocybin-and-ssri-snri-interactions.

76 *Lithium combined with psychedelics*: S. M. Nayak et al., "Classic Psychedelic Coadministration with Lithium, but Not Lamotrigine, Is Associated with Seizures: An Analysis of Online Psychedelic Experience Reports," *Pharmacopsychiatry* 54, no. 5 (2021): 240–45.

77 *So do steroids, calcium channel blockers*: Here's another place to find lists of interactions: "Alternative Treatments for Cluster Headaches," Clusterbusters, https://clusterbusters.org/resource/alternative-treatments-for-cluster-headaches.

77 *The efficacy of the anticoagulant*: A. Nadkarni et al., "Drug-Drug Interactions Between Warfarin and Psychotropics: Updated Review of the Literature," *Pharmacotherapy* 32, no. 10 (2012): 932–42.

77 *In general, interactions*: A. A. Joos, "Pharmakologische Interaktionen von Antibiotika und Psychopharmaka" [Pharmacologic interactions of antibiotics and psychotropic drugs], *Psychiatrische Praxis* 25, no. 2 (1998): 57–60.

77 *though taking them together*: Maria Holyanova, "Can You Mix Psychedelics and Antibiotics? Here's What the Research Says," Psychedelic Spotlight, April 20, 2023, https://psychedelicspotlight.com/can-you-take-psychedelics-while-on-antibiotics-heres-what-the-research-says-with-examples.

77 *Whether it is okay to take mushrooms*: "Psilocybin Mushrooms ('Magic Mushrooms')," Organization of Teratology Information Specialists, August 1, 2021, https://www.ncbi.nlm.nih.gov/books/NBK582810.

77 *older people may experience less*: K. Ko et al., "Predicting the Intensity of Psychedelic-Induced Mystical and Challenging Experience in a Healthy Population: An Exploratory Post-Hoc Analysis," *Neuropsychiatry Disease and Treatment* 19 (2023): 2105–13.

77 *also can produce unpredictable results*: Kim Rose-Francis, "The Effects

of Mixing Mushrooms and Alcohol," Medical News Today, April 29, 2022, https://www.medicalnewstoday.com/articles/the-effects-of-mixing-mushrooms-and-alcohol.

77 *potentially intensifying the effects*: I. Ibarra-Lecue et al., "Chronic Cannabis Promotes Pro-Hallucinogenic Signaling of 5-HT2A Receptors Through Akt/mTOR Pathway," *Neuropsychopharmacology* 43, no. 10 (2018): 2028–35, https://doi.org/10.1038/s41386-018-0076-y.

78 *drug's interactions with other conditions*: Rachel Clark (the Psychedelic Auntie—love her), "How Do I Know If Psychedelics Are Safe to Combine with My Medication?," Lucid News, February 17, 2023, https://www.lucid.news/psychedelics-safe-to-combine-with-medication.

78 *Addiction isn't really considered an issue*: A. K. Schlag et al., "Adverse Effects of Psychedelics: From Anecdotes and Misinformation to Systematic Science," *Journal of Psychopharmacology* 36, no. 3 (2022): 258–72.

78 *HPPD, or hallucinogen-persisting perception disorder*: L. Hermle et al., "Hallucinogen-Persisting Perception Disorder," *Therapeutic Advances in Psychopharmacology* 2, no. 5 (2012): 199–205.

78 *HPPD can reoccur for months*: M. L. Espiard, "Hallucinogen Persisting Perception Disorder After Psilocybin Consumption: A Case Study," *European Psychiatry* 20, no. 5–6 (2005): 458–60.

78 *start after just one trip*: M. Kurtom, A. Henning, and E. D. Espiridion, "Hallucinogen-Persisting Perception Disorder in a 21-Year-Old Man," *Cureus* 11, no. 2 (2019): e4077.

78 *others find it debilitating*: F. Müller et al., "Flashback Phenomena After Administration of LSD and Psilocybin in Controlled Studies with Healthy Participants," *Psychopharmacology (Berlin)* 239, no. 6 (2022): 1933–43.

78 *Antipsychotic medications could make*: Brit Dawson, "This Is What It's Like to Experience Never-Ending Acid Trips," *Face*, August 9, 2022, https://theface.com/life/never-ending-trips-psychedelic-drug-tiktok-mushrooms-acid.

79 *Many people have suggested the culprit*: Barbara Bauer, "Why Do Magic Mushrooms Cause Nausea?," Psychedelic Science Review, August 30, 2019, https://psychedelicreview.com/why-do-magic-mushrooms-cause-nausea.

79 *Even in clinical settings*: Katie MacBride, "Magic Mushrooms: One Weird Trick Could Make Psilocybin Therapy Better," Inverse, May 20, 2021, https://www.inverse.com/mind-body/make-psilocybin-therapy-better.

81 *highly influenced by their environment*: Tim Read and Maria Papaspyrou, eds., *Psychedelics and Psychotherapy: The Healing Potential of Expanded States* (Rochester, VT: Park Street Press, 2021).

82 *"The higher your dose"*: Rachel Clark, "What Do I Need to Know About Tripping Alone in Nature?," Lucid News, August 10, 2021, https://www.lucid.news/what-do-i-need-to-know-about-tripping-alone-in-nature.

83 *Music contributes to the setting*: M. Kaelen et al., "The Hidden Therapist: Evidence for a Central Role of Music in Psychedelic Therapy," *Psychopharmacology (Berlin)* 235, no. 2 (2018): 505–19.

83 *The one at Johns Hopkins*: For a nice summary of the playlist, see: Marc Shapiro, "Inside the Johns Hopkins Psilocybin Playlist," *Dome*, November/December 2020.

83 *"Sensitivity to the therapeutic potential"*: William A. Richards, *Sacred Knowledge: Psychedelics and Religious Experiences* (New York: Columbia University Press, 2016), 189.

84 *Dr. Fadiman's advice*: James Fadiman, *The Psychedelic Explorer's Guide: Safe, Therapeutic, and Sacred Journeys* (Rochester, VT: Park Street Press, 2011), 19.

85 *tryptamines (psilocybin, psilocin, and the other)*: And that's supported here: K. Gotvaldová et al., "Stability of Psilocybin and Its Four Analogs in the Biomass of the Psychotropic Mushroom *Psilocybe cubensis*," *Drug Testing and Analysis* 13, no. 2 (2012): 439–46.

86 *metallic, earthy, bitter*: Anna Wilcox, "What Do Shrooms Taste Like?," Double Blind, September 7, 2021, https://doubleblindmag.com/what-do-shrooms-taste-like.

86 *described liberty caps as "greasy"*: Andy Letcher, *Shroom: A Cultural History of the Magic Mushroom* (London: Faber and Faber, 2006), 16.

86 *teas are only "good"*: Paul Stamets, *Psilocybin Mushrooms of the World* (Berkeley, CA: Ten Speed Press, 1996), 50.

86 *a paper in 2020 found*: Gotvaldová, "Stability of Psilocybin."

87 *injecting the tea intravenously*: Nicholas B. Giancola et al., "A 'Trip' to the Intensive Care Unit: An Intravenous Injection of Psilocybin," *Journal of the Academy of Consultation-Liaison Psychiatry* 62, no. 3 (2021): 370–71.

87 *Cacao is a potential MAOI*: Emma Stone, "Mushroom Chocolate: The New Wave in Mind Altering Edibles," Psychedelic Spotlight, August 1, 2022, https://psychedelicspotlight.com/mushroom-chocolate-the-new-wave-in-mind-altering-edibles/.

5: Foraging

95 *there are maybe 200 species*: A list of psilocybin-containing mushrooms can be found here: "List of Psilocybin Mushrooms," Bionity, https://www.bionity.com/en/encyclopedia/List_of_Psilocybin_mushrooms.html.

96 *most species live in the neotropics*: Anya Ermakova, "Psychoactive Mushrooms in Mexico: Overview of Ecology and Ethnomycology," Chac-

runa Institute, November 12, 2021, https://chacruna.net/psychoactive-mushrooms-in-mexico-overview-of-ecology-and-ethnomycology.

98 *They are considered moderately active*: Paul Stamets, *Psilocybin Mushrooms of the World: An Identification Guide* (Berkeley, CA: Ten Speed Press, 1996), 142–45.

98 *psilocybin has been detected*: "*Psilocybe semilanceata* (Fr.) P.Kumm., 1871," Global Biodiversity Information Facility, https://www.gbif.org/species/144094035.

98 *The fungus feeds on the dead roots*: Stamets, *Psilocybin Mushrooms*, 24.

99 *wrote a piece in the student paper*: Andy Letcher, *Shroom: A Cultural History of the Magic Mushroom* (London: Faber and Faber, 2006), 221.

100 *Too late, unfortunately*: Buck McAdoo, "Mushroom of the Month: *Galerina marginata*," *MushRumors: The Newsletter of the Northwest Mushroomers Association* 26, no. 5 (November 2015): 9.

101 *That's because cyan*: "Why Are Scrubs Usually Green or Blue?," Board Vitals, November 14, 2019, https://www.boardvitals.com/blog/why-scrubs-usually-blue-green.

101 *"If a gilled mushroom"*: Stamets, *Psilocybin Mushrooms*, 53.

101 *unstable psilocin molecules*: Katrina Krämer, "Mystery of Why Magic Mushrooms Go Blue Solved," Chemistry World, December 10, 2019, https://www.chemistryworld.com/news/mystery-of-why-magic-mushrooms-go-blue-solved/4010870.article.

6: Cultivating

112 *One teaspoon of healthy soil*: Merlin Sheldrake, "Why the Hidden World of Fungi Is Essential to Life on Earth," *Guardian*, October 10, 2020.

113 *That cell then divides*: Brian A. Perry, "Are Mushrooms Genetic Individuals or Genetic Mosaics?," *Mycena News*, December 2007.

113 *The French mycologist*: Yachaj, "Mushroom Cultivation: From Falconer to Fanaticus and Beyond," Vaults of Erowid, February 21, 2013, https://erowid.org/plants/mushrooms/mushrooms_article10.shtml.

114 *synthetic psilocybin was rare*: Andrew Weil, "The Use of Psychoactive Mushrooms in the Pacific Northwest: An Ethnopharmacologic Report," *Botanical Museum Leaflets, Harvard University* 25, no. 5 (1977): 135.

115 *mushroom mycelium may be dispersed*: Peter J. A. Shaw and Geoffrey Kibby, "Aliens in the Flowerbeds: The Fungal Biodiversity of Ornamental Woodchips," *Field Mycology* 2, no. 1 (2001): 6–11.

117 *You will essentially be propagating*: Jonathan Ott, *Hallucinogenic Plants of North America* (San Francisco: Wingbow Press, 1976).

119 *It sold over one hundred thousand*: Terence McKenna, *Food of the Gods: The Search for the Original Tree of Knowledge—A Radical History of Plants, Drugs, and Human Evolution* (New York: Bantam, 1992), 243.

119 *in 1983, the mycologist*: Yachaj, "Mushroom Cultivation."

121 *name translates*: John W. Allen's text on San Isidro de Labrador is here: "SAN ISIDRO de LABRADOR—Psilocybe Cubensis," ResearchGate, https://www.researchgate.net/publication/305380833_SAN_ISIDRO_de_LABRADOR_-_Psilocybe_Cubensis.

122 *the tissue of the mushroom*: From the FreshCap mushroom farm blog (love these folks): "How to Clone Mushrooms," FreshCap, https://learn.freshcap.com/growing/how-to-clone-mushrooms.

123 *Increased psychoactivity*: Justin Cooke, "Magic Mushroom Strain Guide (100+ Strains Explained)," Tripsitter, November 2, 2023, https://tripsitter.com/magic-mushrooms/strains.

124 *there can be a pretty big difference*: Oakland Hyphae is an excellent way to stay on top of cultivar potencies: https://www.oaklandhyphae510.com.

124 *Between variety among specimens*: Barbara Bauer, "Chemical Composition Variability in Magic Mushrooms," Psychedelic Science Review, March 4, 2019, https://psychedelicreview.com/chemical-composition-variability-in-magic-mushrooms.

126 *about what it takes to process*: For canning info generally and mushroom cook times especially: "Selecting, Preparing and Canning Vegetables," National Center for Home Food Preservation, https://nchfp.uga.edu/how/can_04/mushrooms.html.

129 *"a century ago in NetYears"*: For a historical snapshot, see this Erowid article: "Alt.drugs FAQ," Vaults of Erowid, https://erowid.org/psychoactives/faqs/faq_alt_drugs.shtml.

130 *"downloaded, printed, distributed"*: Maybe that's an overstatement from a fan; I don't know. But the psychoactive mushroom cultivation scene is quietly gigantic. "Psylocybe Fanaticus/Robert 'Billy' Mcpherson R.I.P.," Shroomery forum, November 17, 2011, https://www.shroomery.org/forums/showflat.php/Number/15382555/fpart/all.

130 *The alt. hierarchy was founded*: This history from a few places: Patrick Howell O'Neill, "How the Internet Powered a DIY Drug Revolution," Daily Dot, updated June 1, 2021, https://www.dailydot.com/unclick/hive-silk-road-online-drug-culture-history; Jacqueline L. Schneider, "Hiding in Plain Sight: An Exploration of the Illegal(?) Activities of a Drugs Newsgroup," *Howard Journal of Criminal Justice* 42, no. 4 (2003); and Robert McPherson's own memoir, "The United States v. Psylocybe Fanaticus," https://www.seanet.com/~rwmcpherson/pfcase.html.

132 *"masterpiece . . . endorsed by spiritual healers"*: Maria Holyanova, "Mycologist Creates Hybrid Magic Mushroom Strain Called Sacred Sun," Psychedelic Spotlight, March 29, 2023, https://psychedelicspotlight.com/mycologist-creates-hybrid-magic-mushroom-strain-sacred-sun.

7: Clinical Trials, Retreats, and Private Guides

135 *"So now . . . we have a growing demand"*: James Kent, "Psychedelic Clinical Trials and the Michael Pollan Effect," Psychedelic Spotlight, August 9, 2022, https://psychedelicspotlight.com/psychedelic-clinical-trials-and-the-michael-pollan-effect.

135 *only one psilocybin study*: Track it down here: "Efficacy, Safety, and Tolerability of COMP360 in Participants with TRD," ClinicalTrials.gov, https://classic.clinicaltrials.gov/ct2/show/results/NCT05624268.

138 *"How easy is it for* you *to be held?"*: Aixalà was interviewed by the excellent Jules Evans on *Psychedelic Integration*: "Marc Aixala on Psychedelic Integration," Jules Evans, November 10, 2022, YouTube video, 50:23, https://www.youtube.com/watch?v=z2NPG1mBfv8.

138 *a scientist's objectivity*: B. Kious, Z. Schwartz, and B. Lewis, "Should We Be Leery of Being Leary? Concerns About Psychedelic Use by Psychedelic Researchers," *Journal of Psychopharmacology* 37, no. 1 (2023): 45–48.

138 *intimate knowledge of the drug*: Shayla Love, "The Ethics of Taking the Drugs You Study," *Vice*, May 14, 2019, https://www.vice.com/en/article/qv7jkx/the-ethics-of-taking-the-drugs-you-study.

139 *"The placebo effect"*: Ted J. Kaptchuk, "No Better Than a Placebo," *New York Times*, October 10, 2023.

139 *patient with depression*: Natalie Gukasyan, "On Blinding and Suicide Risk in a Recent Trial of Psilocybin-Assisted Therapy for Treatment-Resistant Depression," *Med* 4, no. 1 (2023): 8–9.

139 *study participants who receive a placebo*: David Yaden, Brian Earp, and Roland Griffiths, "Ethical Issues Regarding Nonsubjective Psychedelics as Standard of Care," *Cambridge Quarterly of Healthcare Ethics* 31, no. 4 (2022).

139 *going to compare the efficacy*: Robin Carhart-Harris and Guy M. Goodwin, "The Therapeutic Potential of Psychedelic Drugs: Past, Present, and Future," *Neuropsychopharmacology* 42, no. 11 (2017): 2105–13.

139 *Researchers screen carefully*: C. Bree Johnston et al., "The Safety and Efficacy of Psychedelic-Assisted Therapies for Older Adults: Knowns and Unknowns," *American Journal of Geriatric Psychiatry* 31, no. 1 (2023): 44–53.

141 *Johns Hopkins conducted a survey*: T. M. Carbonaro et al., "Survey Study of Challenging Experiences After Ingesting Psilocybin Mushrooms: Acute and Enduring Positive and Negative Consequences," *Journal of Psychopharmacology* 30, no. 12 (2016): 1268–78.

142 *Psychedelics are more catalysts*: William A. Richards, *Sacred Knowledge: Psychedelics and Religious Experiences* (New York: Columbia University Press, 2016), 119.

143 *Dutch government banned the mushrooms*: Andy Letcher, *Shroom: A Cultural History of the Magic Mushroom* (London: Faber and Faber, 2006), 282.

144 *study in 2023 suggested the dampening effect*: N. Gukasyan et al., "Attenuation of Psilocybin Mushroom Effects During and After SSRI/SNRI Antidepressant Use," *Journal of Psychopharmacology* 37, no. 7 (2023): 707–16.

147 *first psilocybin service center*: Andrew Selsky, "Oregon Launches Legal Psilocybin Access Amid High Demand and Hopes for Improved Mental Health Care," Associated Press, September 15, 2023, https://apnews.com/article/psilocybin-oregon-magic-mushrooms-psychedelics-therapy-legal-6e5389b090b0c50d5c90d9574b63eca5.

147 *Psychedelic retreats are not cheap*: "Oregon Opens the First Facility to Consume Legal Mushrooms," Young Turks, May 13, 2023, YouTube video, 6:30, https://www.youtube.com/watch?v=bc3TQxGITKk.

148 *"Just as a single calamitous incident"*: James Fadiman, *The Psychedelic Explorer's Guide: Safe, Therapeutic, and Sacred Journeys* (Rochester, VT: Park Street Press, 2011), 113.

149 *"don't want to go to Peru"*: Brittany Chang, "A Canadian Company Is Launching a Luxury Psychedelic Retreat with Private Cabins, Psilocybin, and Cannabis," *Business Insider*, July 30, 2022, https://www.businessinsider.com/photos-canadian-company-launching-luxury-psychedelic-marijuana-psilocybin-cacao-retreat-2022-7.

150 *According to Haberstroh*: *Psychedelics Today* podcast, episode 244, "Mark Haberstroh—Mushrooms, Retreat Centers, and Safety," May 18, 2021, https://psychedelicstoday.com/2021/05/18/pt244-mark-haberstroh-mushrooms-retreat-centers-and-safety.

154 *its reputation took a major hit*: *New York* magazine's podcast *Cover Story: Power Trip* is a chilling report on these abuses.

155 *therapist might try to separate*: Olivia Goldhill, "A Psychedelic Therapist Allegedly Took Millions from a Holocaust Survivor, Highlighting Worries About Elders Taking Hallucinogens," Stat, April 21, 2022, https://www.statnews.com/2022/04/21/psychedelic-therapist-allegedly-took-millions-from-holocaust-survivor-highlighting-worries-about-elders-taking-hallucinogens.

163 *review published*: Alexander Trope et al., "Psychedelic-Assisted Group Therapy: A Systematic Review," *Journal of Psychoactive Drugs* 51, no. 2 (2019): 174–88.

164 *"We are the experts of ourselves"*: Marc Aixalà on the *Psychedelics Today* podcast, episode 374, "Personalizing Psychedelic Integration," November 22, 2022, https://psychedelicstoday.com/2022/11/22/pt374.

166 *"My teachers"*: *EntheoNation* podcast, episode 29, "Experimental Mushroom Shamanism & Mediumship," https://entheonation.com/episodes/magic-mushroom-shonagh-home.

166 *Neoshamans tend to be self-appointed*: J. Scuro and R. Rodd, "Neo-Shamanism," in *Encyclopedia of Latin American Religions*, ed. H. Gooren (Cham, Switzerland: Springer, 2019), https://doi.org/10.1007/978-3-319-08956-0_49-1.

167 *"and instead, use anecdotes"*: Patric Plesa and Rotem Petranker, "Manifest Your Desires: Psychedelics and the Self-Help Industry," *International Journal of Drug Policy* 105 (2022), article 103704.

8: Microdosing

173 *2021 Global Drug Survey*: Global Drug Survey 2021, https://www.globaldrugsurvey.com/wp-content/uploads/2021/12/Report2021_global.pdf.

175 *many microdosers consume*: Megan Webb, Heith Copes, and Peter S. Hendricks, "Narrative Identity, Rationality, and Microdosing Classic Psychedelics," *International Journal of Drug Policy* 70 (2019): 33–39.

175 *A Google search for microdosing*: B. Beaton et al., "Accounting for Microdosing Classic Psychedelics," *Journal of Drug Issues* 50, no. 1 (2020): 3–14.

175 *trend data showed*: To stay atop of trend data, see: "Keyword Overview: Microdosing," Ubersuggest, https://app.neilpatel.com/en/ubersuggest/overview?lang=en&locId=2840&keyword=Microdosing.

176 *articles in* Rolling Stone *and* Wired *described*: Andrew Leonard, "How LSD Microdosing Became the Hot New Business Trip," *Rolling Stone*, November 20, 2015, https://www.rollingstone.com/culture/culture-news/how-lsd-microdosing-became-the-hot-new-business-trip-64961; Olivia Solon, "Under Pressure, Silicon Valley Workers Turn to LSD Microdosing," *Wired (UK)*, August 24, 2016, https://www.wired.co.uk/article/lsd-microdosing-drugs-silicon-valley.

176 *In a small Dutch study*: L. Prochazkova et al., "Exploring the Effect of Microdosing Psychedelics on Creativity in an Open-Label Natural Setting," *Psychopharmacology (Berlin)*, 235, no. 12 (2018): 3401–13.

177 *"we believe that creativity"*: Shane Heath, "Why Should Employees Have to Keep Microdosing a Secret?," MUD\WTR, March 24, 2022, https://mudwtr.com/blogs/trends-with-benefits/why-mud-wtr-allows-microdosing-at-work?sscid=91k6_mnfvw&.

177 *there is some evidence*: Eline C. H. M. Haijen, Petra P. M. Hurks, and Kim P. C. Kuypers, "Microdosing with Psychedelics to Self-Medicate for ADHD Symptoms in Adults: A Prospective Naturalistic Study," *Neuroscience Applied* 1 (2022), article 101012.

177 *Ayelet Waldman's book*: Ayelet Waldman, *A Really Good Day: How Microdosing Made a Mega Difference in My Mood, My Marriage, and My Life* (New York: Anchor Books, 2018).

178 *studies have shown that microdosing*: K. F. Kiilerich et al., "Repeated Low Doses of Psilocybin Increase Resilience to Stress, Lower Compulsive Actions, and Strengthen Cortical Connections to the Paraventricular Thalamic Nucleus in Rats," *Molecular Psychiatry* 28 (2023): 3829–41, https://doi.org/10.1038/s41380-023-02280-z.

178 *adults who microdose*: J. M. Rootman et al., "Adults Who Microdose Psychedelics Report Health Related Motivations and Lower Levels of Anxiety and Depression Compared to Non-Microdosers," *Scientific Reports* 11 (2021), article 22479, https://doi.org/10.1038/s41598-021-01811-4.

178 P. cubensis *in general contains*: Paul Stamets, *Psilocybin Mushrooms of the World: An Identification Guide* (Berkeley, CA: Ten Speed Press, 1996), 39.

178 *review study in 2020*: K. P. C. Kuypers, "The Therapeutic Potential of Microdosing Psychedelics in Depression," *Therapeutic Advances in Psychopharmacology* 10 (2020), https://doi.org/10.1177/2045125320950567.

179 *online self-reporting forums*: Rootman et al., "Adults Who Microdose"; T. Anderson et al., "Microdosing Psychedelics: Personality, Mental Health, and Creativity Differences in Microdosers," *Psychopharmacology (Berlin)* 236, no. 2 (2019): 731–40.

179 *study conducted by Dr. Fadiman*: James Fadiman and Sophia Korb, "Might Microdosing Psychedelics Be Safe and Beneficial? An Initial Exploration," *Journal of Psychoactive Drugs* 51, no. 2 (2019): 118–22.

179 *have not shown that microdosing*: B. Szigeti et al., "Self-Blinding Citizen Science to Explore Psychedelic Microdosing," *eLife* 2, no. 10 (2021): e62878, https://doi.org/10.7554/eLife.62878.

179 *any mood-related differences*: F. Cavanna et al., "Microdosing with Psilocybin Mushrooms: A Double-Blind Placebo-Controlled Study," *Translational Psychiatry* 12 (2022), article 307, https://doi.org/10.1038/s41398-022-02039-0.

179 *main reason why people stop*: N. R. P. W. Hutten et al., "Motives and Side-Effects of Microdosing with Psychedelics Among Users," *International Journal of Neuropsychopharmacology* 22, no. 7 (2019): 426–34.

179 *a researcher in how altered states can affect*: Vince Polito and R. J. Stevenson, "A Systematic Study of Microdosing Psychedelics," *PLOS ONE* 14, no. 2 (2019): e0211023, https://doi.org/10.1371/journal.pone.0211023.

180 *Apex Labs received approval*: Ben Hargreaves, "Apex Gets Approval for 'World's Largest' Take Home Psychedelic Trial," Outsourcing-Pharma, January 23, 2023, https://www.outsourcing-pharma.com/Article/2023/01/23/apex-gets-approval-for-biggest-take-home-psychedelic-trial.

180 *It turns out his wife*: James Hallifax, "Senator's Wife Secretly Gave Him

Psilocybin to Alleviate Depression," Psychedelic Spotlight, May 21, 2022, https://psychedelicspotlight.com/canadian-senators-wife-secretly-gave-him-psilocybin-to-help-his-depression.

183 *Child Protective Services takes away*: "Racially Disproportionate Drug Arrests," in *Punishment and Prejudice: Racial Disparities in the War on Drugs*, Human Rights Watch, https://www.hrw.org/reports/2000/usa/Rcedrg00-05.htm.

183 *"I think it's really important"*: May Richard-Craven, "The Black Mothers Finding Freedom in Mushrooms: 'They Give Us Our Power Back,'" *Guardian*, August 28, 2022, https://www.theguardian.com/society/2022/aug/28/mushrooms-psychedelic-black-women-mothers-microdosing.

184 *Even* Elle *magazine*: Kelly Mickle, "Shrooms Are the New Cali Sober," *Elle*, April 12, 2023, https://www.elle.com/culture/a43469022/shrooms-are-the-new-cali-sober-april-2023.

185 *risk hasn't been comprehensively assessed*: M. Tagen et al., "The Risk of Chronic Psychedelic and MDMA Microdosing for Valvular Heart Disease," *Journal of Psychopharmacology* 37, no. 9 (2023): 876–90.

185 *microdosers have reported*: All about side effects: James Fadiman, "Microdose Research: Without Approvals, Control Groups, Double-Blinds, Staff or Funding," Psychedelic Press, November 16, 2017, https://psychedelicpress.co.uk/blogs/psychedelic-press-blog/microdose-research-james-fadiman; Hutten et al., "Motives and Side-Effects of Microdosing"; and T. Anderson et al., "Psychedelic Microdosing Benefits and Challenges: An Empirical Codebook," *Harm Reduction Journal* 16 (2019), article 43.

185 *exacerbate some conditions*: Petter Grahl Johnstad, "Powerful Substances in Tiny Amounts: An Interview Study of Psychedelic Microdosing," *Nordic Studies on Alcohol and Drugs* 35, no. 1 (2018): 39–51.

185 *microdosing seems to be well tolerated*: Kuypers, "The Therapeutic Potential of Microdosing."

186 *taking SSRIs appears to weaken the effects*: A. M. Becker et al., "Acute Effects of Psilocybin After Escitalopram or Placebo Pretreatment in a Randomized, Double-Blind, Placebo-Controlled, Crossover Study in Healthy Subjects," *Clinical Pharmacology & Therapeutics* 111, no. 4 (2022): 886–95, https://doi.org/10.1002/cpt.2487.

186 *microdosers take mushrooms for anywhere*: K. P. C. Kuypers et al., "Microdosing Psychedelics: More Questions Than Answers? An Overview and Suggestions for Future Research," *Journal of Psychopharmacology* 33, no. 9 (2019): 1039–57, https://doi.org/10.1177/0269881119857204.

186 *primary concern people have*: Anderson, "Psychedelic Microdosing Benefits and Challenges."

186 *Take the case of Jessica Thornton*: Emma Stone, "Indiana Nurse Faces Ten Years in Prison for Microdosing Psilocybin Mushrooms," Psychedelic Spotlight, July 8, 2022, https://psychedelicspotlight.com/indiana-nurse-faces-ten-years-in-prison-for-microdosing-psilocybin-mushrooms. I learned the update on her case from Aaron Paul Orsini.

190 *what users call*: Jaden Rae, *Microdosing Guide and Journal: A Complete Step-by-Step Microdose Guide to Safely Source, Track & Customize Your Medicinal Mushroom Journey* (self-published, Fresh Karma, 2022), 29.

191 *A high dose*: Albert Garcia-Romeu et al., "Optimal Dosing for Psilocybin Pharmacotherapy: Considering Weight-Adjusted and Fixed Dosing Approaches," *Journal of Psychopharmacology* 35, no. 4 (2021): 353–61.

191 *Many folks don't even know*: Hutten et al., "Motives and Side-Effects of Microdosing."

192 *According to an analysis*: M. Pellegrini et al., "Magic Truffles or Philosopher's Stones: A Legal Way to Sell Psilocybin?," *Drug Testing and Analysis* 5, no. 3 (2013): 182–85.

193 *recreational dose of fresh truffles*: Prochazkova et al., "Exploring the Effect of Microdosing."

197 *analysis of their microdosing habits*: Rootman et al., "Adults Who Microdose."

199 *time perception researcher*: Marc Wittmann et al., "Effects of Psilocybin on Time Perception and Temporal Control of Behaviour in Humans," *Journal of Psychopharmacology* 21, no. 1 (2007): 50–64.

9: Therapeutic Trips

201 *They are financing studies*: "Compass Pathways Announces Positive Topline Results from Groundbreaking Phase 2b Trial of Investigational COMP360 Psilocybin Therapy for Treatment-Resistant Depression," Compass Pathways, November 9, 2021, https://compasspathways.com/positive-topline-results.

204 *Kreitman was part of a pilot study*: Matthew W. Johnson et al., "Pilot Study of the 5-HT2AR Agonist Psilocybin in the Treatment of Tobacco Addiction," *Journal of Psychopharmacology* 28, no. 11 (2014): 983–92; Matthew W. Johnson, Albert Garcia-Romeu, and Roland R. Griffiths, "Long-Term Follow-Up of Psilocybin-Facilitated Smoking Cessation," *American Journal of Drug and Alcohol Abuse* 43, no. 1 (2017): 55–56.

204 *In a 2022 randomized double-blind study*: M. P. Bogenschutz et al., "Percentage of Heavy Drinking Days Following Psilocybin-Assisted Psychotherapy vs Placebo in the Treatment of Adult Patients with Alcohol Use Disorder: A Randomized Clinical Trial," *JAMA Psychiatry* 79, no. 10 (2022): 953–62.

208 *Anorexia nervosa is a life-threatening*: Sarah-Catherine Rodan et al., "Psilocybin as a Novel Pharmacotherapy for Treatment-Refractory Anorexia Nervosa," *OBM Neurobiology* 5, no. 2 (2021): 1–25.

208 *Psilocybin may help*: M. J. Spriggs, Hannes Kettner, and Robin Carhart-Harris, "Positive Effects of Psychedelics on Depression and Wellbeing Scores in Individuals Reporting an Eating Disorder," *Eating and Weight Disorders* 26, no. 4 (2021): 1265–70.

208 *"The question is not so much"*: Marianne Apostolides, "Psychedelics Offer New Route to Recovery from Eating Disorders," proto.life, February 3, 2022, https://proto.life/2022/02/psychedelics-offer-new-route-to-recovery-from-eating-disorders.

209 *more vets have died from suicide*: Jennifer Steinhauer, "Suicides Among Post-9/11 Veterans Are Four Times as High as Combat Deaths," *New York Times*, June 22, 2021.

210 *efficacy of psilocybin in treating PTSD*: Ernesto Londoño, "After Six-Decade Hiatus, Experimental Psychedelic Therapy Returns to the V.A.," *New York Times*, June 24, 2022; Feleshia Chandler, "N.S. Company Launching Clinical Trial to Examine Magic Mushrooms as Treatment for PTSD," CBC News, October 24, 2022, https://www.cbc.ca/news/canada/nova-scotia/psilocybin-research-windsor-ns-ptsd-1.6615289.

210 *complementary study in Europe*: K. N. Elsouri et al., "Psychoactive Drugs in the Management of Post Traumatic Stress Disorder: A Promising New Horizon," *Cureus* 14, no. 5 (2022): e25235.

211 *transition of acute pain*: Joel P. Castellanos et al., "Chronic Pain and Psychedelics: A Review and Proposed Mechanism of Action," *Regional Anesthesia & Pain Medicine* 45, no. 7 (2020): 486–94; Clare Watson, "The Psychedelic Remedy for Chronic Pain," *Nature*, September 28, 2022, https://www.nature.com/articles/d41586-022-02878-3.

211 *psilocybin may reset*: S. M. Nkadimeng, C. M. L. Steinmann, and J. N. Eloff, "Anti-Inflammatory Effects of Four Psilocybin-Containing Magic Mushroom Water Extracts in Vitro on 15-Lipoxygenase Activity and on Lipopolysaccharide-Induced Cyclooxygenase-2 and Inflammatory Cytokines in Human U937 Macrophage Cells," *Journal of Inflammation Research* 14 (2021): 3729–38.

212 *enduring effect on migraine headache*: Emmanuelle Schindler et al., "Exploratory Controlled Study of the Migraine-Suppressing Effects of Psilocybin," *Neurotherapeutics* 18, no. 1 (2021): 534–43.

213 *one man actually did shoot himself*: "Emmanuelle Schindler: Indoleamine Hallucinogens in Treatment of Cluster Headache," MAPS, May 10, 2017, YouTube video, 31:30, https://www.youtube.com/watch?v=GmpP_n86 9c.

213 *which anecdotally showed psilocybin*: R. Andrew Sewell, John H. Halpern, and Harrison G. Pope Jr., "Response of Cluster Headache to Psilocybin and LSD," *Neurology* 66, no. 12 (2006): 1920–22.

215 *may be as effective as the SSRI Lexapro*: B. Weiss et al., "Personality Change in a Trial of Psilocybin Therapy v. Escitalopram Treatment for Depression," *Psychological Medicine* 54, no. 1 (2023): 178–92.

215 *Compass study looked at people*: G. M. Goodwin et al., "Single-Dose Psilocybin for a Treatment-Resistant Episode of Major Depression," *New England Journal of Medicine* 387, no. 18 (2022): 1637–48.

217 *training is key*: "Psychedelics: A Journey Through Grief and Loss," End Well, November 13, 2021, YouTube video, 46:03, https://www.youtube.com/watch?v=XZ3WM12H2Fs.

218 *PTSD is understood as resulting*: "Healing Your Brain After Loss," American Brain Foundation, June 24, 2021, YouTube video, 58:49, https://www.youtube.com/watch?v=hZwhslOz7qY.

220 *rites around death*: A. Santarpia et al., "The Narrative Effects of Shamanic Mythology in Palliative Care," *Journal of Humanistic Psychology* 61, no. 1 (2021): 73–103.

221 *it was the mystical experience*: Roland R. Griffiths et al., "Psilocybin-Occasioned Mystical-Type Experience in Combination with Meditation and Other Spiritual Practices Produces Enduring Positive Changes in Psychological Functioning and in Trait Measures of Prosocial Attitudes and Behaviors," *Journal of Psychopharmacology* 32, no. 1 (2018): 49–69; B. Kelmendi et al., "The Role of Psychedelics in Palliative Care Reconsidered: A Case for Psilocybin," *Journal of Psychopharmacology* 30, no. 12 (2016): 1212–14.

221 *"so completely is it set"*: Huston Smith, *Cleansing the Doors of Perception: The Religious Significance of Entheogenic Plants and Chemicals* (New York: Penguin Putnam, 2000), 142.

10: Spiritual Trips

229 *"I see that I am a link"*: James Fadiman, *The Psychedelic Explorer's Guide: Safe, Therapeutic, and Sacred Journeys* (Rochester, VT: Park Street Press, 2011), 42.

229 *called* nature relatedness: William A. Richards, *Sacred Knowledge: Psychedelics and Religious Experiences* (New York: Columbia University Press, 2016), 19.

229 *"From Egoism to Ecoism"*: Hannes Kettner et al., "From Egoism to Ecoism: Psychedelics Increase Nature Relatedness in a State-Mediated and Context-Dependent Manner," *International Journal of Environmental Research and Public Health* 16, no. 24 (2019): 5147.

230 *term was coined*: Carl A. P. Ruck et al., "Entheogens," *Journal of Psychedelic Drugs* 11, no. 1–2 (1979): 145–46.

230 *Dr. Fadiman quotes a microdoser*: Fadiman, *The Psychedelic Explorer's Guide*, 201.

233 *drug becomes an entheogen*: Andy Letcher, "Mad Thoughts on Mushrooms: Discourse and Power in the Study of Psychedelic Consciousness," *Anthropology of Consciousness* 18, no. 2 (2007): 74–98.

233 *deep spiritual journey is going to unfold*: Fadiman, *The Psychedelic Explorer's Guide*, 30.

234 *"inner authority"*: William James, *The Varieties of Religious Experience: A Study in Human Nature* (New York: Penguin, 1982), 16; Ron Cole-Turner, "Psychedelic Epistemology: William James and the 'Noetic Quality' of Mystical Experience," *Religions* 12, no. 12 (2021): 1058.

234 *there is another reality*: Huston Smith, *Cleansing the Doors of Perception: The Religious Significance of Entheogenic Plants and Chemicals* (New York: Penguin Putnam, 2000), 133.

234 *Brian Muraresku, author of*: "Video: Psychedelics: The Ancient Religion with No Name?," Center for the Study of World Religions, Harvard Divinity School, February 12, 2021, https://cswr.hds.harvard.edu/news/2021/02/12/video-psychedelics-ancient-religion-no-name.

235 *Wasson proposed that ancient humans*: R. Gordon Wasson, "Seeking the Magic Mushroom," *Life*, May 1957, https://bibliography.maps.org/bibliography/default/resource/15048.

235 *proposed by Brian P. Akers*: Brian P. Akers et al., "A Prehistoric Mural in Spain Depicting Neurotropic *Psilocybe* Mushrooms?," *Economic Botany* 65 (2011): 121–28.

235 *the mushrooms represented*: "The Stone Mushrooms of Thrace Conference Proceedings 2012," Giorgio Samorini, https://www.samorini.it/doc1/sam/samorini%20tracia.pdf.

237 *called* Ndi Xijtho: Osiris García Cerqueda, "Magic Mushrooms, Memory and Resistance in the Sierra Mazateca," Chacruna Institute, August 18, 2020, https://chacruna.net/mushrooms-resistance-sierra-mazateca.

237 *Huautla de Jiménez, has hosted a sacred tourism*: Cerqueda, "Magic Mushrooms."

237 *different types of tourists*: Inti García Flores, Rosalía Acosta López, and Sarai Piña Alcántara, "Niños Santos, Psilocybin Mushrooms and the Psychedelic Renaissance," Chacruna Institute, November 12, 2020, https://chacruna.net/mazatec_mushroom_ceremony_psychedelic_tourism.

238 *It is constantly reconfigured*: Cerqueda, "Magic Mushrooms."

238 *the* chjine *empathizes*: Flores, López, and Alcántara, "Niños Santos."

240 *"We are not isolated entities"*: "Horizons Northwest 2022: MARIO

ALONSO MARTÍNEZ CORDERO," HorizonsConference, December 1, 2022, YouTube video, 42:20, https://www.youtube.com/watch?v=ts2Ufx-wYV0.

243 *Marsh Chapel experiment*: Smith, *Cleansing the Doors of Perception*, 99.

244 *a twenty-five-year follow-up*: R. Doblin, "Pahnke's 'Good Friday Experiment': A Long-Term Follow-Up and Methodological Critique," *Journal of Transpersonal Psychology* 23, no. 1 (1991): 1–28.

244 *(The fellow who was sedated)*: Smith, *Cleansing the Doors of Perception*, 104.

244 *the 2006 publication of a blockbuster*: Roland R. Griffiths et al., "Psilocybin Can Occasion Mystical-Type Experiences Having Substantial and Sustained Personal Meaning and Spiritual Significance," *Psychopharmacology (Berlin)* 187, no. 3 (2006): 268–83; discussion 284–92.

245 *One study on the relationship*: S. Kangaslampi, A. Hausen, and T. Rauteenmaa, "Mystical Experiences in Retrospective Reports of First Times Using a Psychedelic in Finland," *Journal of Psychoactive Drugs* 52, no. 4 (2020): 309–18.

245 *ability to decrease self-referential*: Frederick S. Barrett and Roland R. Griffiths, "Classic Hallucinogens and Mystical Experiences: Phenomenology and Neural Correlates," *Current Topics in Behavioral Neurosciences* 36 (2018): 393–430.

245 *"Drugs appear to induce religious experiences"*: Smith, *Cleansing the Doors of Perception*, 30.

245 *study included religious professionals*: The study remains unpublished as of press time, ClinicalTrial.gov ID NCTO2421263.

247 *"There are, of course, numerable drug"*: Smith, *Cleansing the Doors of Perception*, 30.

248 *"Science has to deal with us"*: "Psychedelics and the Seminary Lecture Series 2022," Starr King School for the Ministry, March 29, 2022, https://www.sksm.edu/academics/the-center-for-multi-religious-studies/psychedelics-and-the-seminary-lecture-series.

251 *the Farm, a spiritual commune*: From a May 30, 2016, review by G. William Barnard of Morgan Shipley, *Psychedelic Mysticism: Transforming Consciousness, Religious Experiences, and Voluntary Peasants in Postwar America* (Lanham, MD: Lexington Books, 2015), https://readingreligion.org/9781498509091/psychedelic-mysticism.

254 *Legally, Zide Door takes the position*: Jessica De La Torre, "Oakland's Psychedelic Mushroom Church Makes a Cautious Return," Oaklandside, June 10, 2022, https://oaklandside.org/2022/06/10/zide-door-psycedelic-magic-mushroom-church-oakland.

254 *"the mushroom rabbi"*: Philissa Cramer, "Denver Drops Charges

Against 'Mushroom Rabbi' Preaching Psychedelics," *Denver Post*, December 13, 2022.

255 *"the good of mankind"*: United States v. Meyers, 906 F. Supp. 1494 (D. Wyo. 1995).

255 *church is going to include entheogens*: Christian Greer and Paul Gillis-Smith, "Psychedelics, Spirituality, and a Culture of Seekership," *Harvard Religion Beat* podcast, episode 11, May 12, 2022.

256 *rapid secularization*: Gregory A. Smith, "About Three-in-Ten U.S. Adults Are Now Religiously Unaffiliated," Pew Research Center, December 14, 2021, https://www.pewresearch.org/religion/2021/12/14/about-three-in-ten-u-s-adults-are-now-religiously-unaffiliated.

11: Psychedelic Community

257 *various tripping collectives have flourished*: J. Christian Greer, "The Greening of Psychedelics," *Harvard Divinity Bulletin*, Autumn/Winter 2022.

258 *he and his colleagues explored*: M. Agrawal et al., "Assessment of Psilocybin Therapy for Patients with Cancer and Major Depression Disorder," *JAMA Oncology* 9, no. 6 (2023): 864–66.

258 *Those involved said the group format*: Marc Gunther, "Group Therapy Holds Promise for Psychedelic-Based Treatments," Lucid News, March 8, 2023, https://www.lucid.news/group-therapy-holds-promise-for-psychedelic-based-treatments; Amanda Siebert, "Group Therapy Plays Important Role in New Study on Psilocybin for Depression in Cancer Patients," Dales Report, February 15, 2022, https://thedalesreport.com/interviews/group-therapy-plays-important-role-in-new-study-on-psilocybin-for-depression-in-cancer-patients.

258 *Another study that looked at how psychedelics*: P. S. Hendricks, "Psilocybin-Assisted Group Therapy: A New Hope for Demoralization," *eClinicalMedicine* 27 (2020): 100557.

258 *participants have asked to meet*: Alexander Trope et al., "Psychedelic-Assisted Group Therapy: A Systematic Review," *Journal of Psychoactive Drugs* 51, no. 2 (2019): 174–88.

260 *"Chacruna Debunks 6 Racist Myths"*: "Chacruna Debunks 6 Racist Myths from the Psychedelic Community," Chacruna Institute, February 15, 2022, video, 15:14, https://chacruna.net/chacruna-debunks-6-racist-myths-from-the-psychedelic-community.

266 *marginalized people often fear going public*: "Chacruna Debunks 6 Racist Myths."

266 *she considered dosing Richard Nixon*: Dave Swanson, "How Grace Slick Planned to Dose President Nixon with LSD," Ultimate Classic Rock,

April 24, 2015, https://ultimateclassicrock.com/grace-slick-president-nixon-acid-1970.

267 *"You do it through this process"*: Clay Skipper, "Where the Psychedelic Revolution Is Headed, According to the Guy Who (Arguably) Started It," *GQ*, October 26, 2021, https://www.gq.com/story/rick-doblin-interview-where-the-psychedelic-revolution-is-headed.

267 *"QAnon shaman"*: "Chacruna Debunks 6 Racist Myths."

267 *"non-conformist thinking styles"*: A. V. Lebedev et al., "Alternative Beliefs in Psychedelic Drug Users," *Scientific Reports* 13, no. 1 (2023): 16432.

269 *"holistic healing center"*: Ken Jordan, "Dennis McKenna's Heroic Mushroom Journey," Lucid News, February 23, 2023, https://www.lucid.news/dennis-mckennas-heroic-mushroom-journey.

273 *"spontaneous* communitas": Hannes Kettner et al., "Psychedelic Communitas: Intersubjective Experience During Psychedelic Group Sessions Predicts Enduring Changes in Psychological Wellbeing and Social Connectedness," *Frontiers in Pharmacology* 12 (2021): 623985.

273 *how a rave creates community*: M. Newson et al., "'I Get High with a Little Help from My Friends'—How Raves Can Invoke Identity Fusion and Lasting Co-operation via Transformative Experiences," *Frontiers in Psychology* 12 (2021): 719596.

276 *calling it* grokking: David Luke, *Otherworlds: Psychedelics and Exceptional Human Experience* (London: Aeon Books, 2019), 173.

277 *An early test of ESP*: Shayla Love, "The Long Strange Relationship Between Psychedelics and Telepathy," *Vice*, July 18, 2022, https://www.vice.com/en/article/z34xa5/the-long-strange-relationship-between-psychedelics-and-telepathy.

277 *a few researchers are pursuing it*: David Luke, "Psychoactive Substances and Paranormal Phenomena: A Comprehensive Review," *International Journal of Transpersonal Studies* 31, no. 1 (2012): 97–156.

277 *In a 2008 survey*: David Luke and Marios Kittenis, "A Preliminary Survey of Paranormal Experiences with Psychoactive Drugs," *Journal of Parapsychology* 69, no. 2 (2005): 305–27.

277 *similar findings in 2020*: Petter Grahl Johnstad, "Psychedelic Telepathy: An Interview Study," *Journal of Scientific Exploration* 34, no. 3 (2020): 493.

Selected Bibliography

Note: There are many books on magic mushrooms and related subjects in Spanish, which I don't read. If you do, go to Chacruna's "The 40 Best Books About Shamanism and Plant Medicines" on their website, chacruna.net.

Aixalà, Marc. *Psychedelic Integration: Psychotherapy for Non-Ordinary States of Consciousness.* Santa Fe, NM: Synergetic Press, 2022.

Allegro, John M. *The Sacred Mushroom and the Cross.* New York: Doubleday, 1970.

Arora, David. *Mushrooms Demystified.* Berkeley, CA: Ten Speed Press, 1979.

Barlow, Caine. *Australian Entheogenic Compendium.* Belgrave, Victoria, Australia: Entheogenesis Australis, 2017.

Belser, Alex, Clancy Cavnar, and Beatriz C. Labate, eds. *Queering Psychedelics: From Oppression to Liberation in Psychedelic Medicine.* Santa Fe, NM: Synergetic Press, 2022.

Blanchard, Geral T. *Awakening the Healing Soul: Indigenous Wisdom for Today's World.* Self-published, Center for Peace Research, 2020.

Bone, Eugenia. *Mycophilia: Revelations from the Weird World of Mushrooms.* New York: Rodale, 2011.

Brown, Jerry B., and Julie M. Brown. *The Psychedelic Gospels: The Secret History of Hallucinogens in Christianity.* Rochester, VT: Park Street Press, 2016.

Brown, Robert E., ed. *The Psychedelic Guide to Preparation of the Eucharist in a Few of Its Many Guises*, 4th ed. Austin: Linga Sharia Incense Co., 1975.

Bunyard, Britt A. *The Lives of Fungi: A Natural History of Our Planet's Decomposers.* Princeton, NJ: Princeton University Press, 2022.

Bunyard, Britt A., and Jay Justice. *Amanitas of North America.* Batavia, IL: Fungi Press, 2020.

Campbell, Joseph. *The Hero with a Thousand Faces.* San Francisco: New World Library, 2008.

Carpenter, Dan. *A Psychonaut's Guide to the Invisible Landscape: The Topography of the Psychedelic Experience.* Rochester, VT: Park Street Press, 2006.

Fadiman, James. *The Psychedelic Explorer's Guide: Safe, Therapeutic, and Sacred Journeys.* Rochester, VT: Park Street Press, 2011.

Grof, Stanislav. *The Way of the Psychonaut: Encyclopedia for Inner Journeys*, vols. 1–2. San Jose, CA: MAPS, 2019.

Guzmán, Gastón. *The Genus* Psilocybe*: A Systematic Revision of the Known Species Including the History, Distribution and Chemistry of the Hallucinogenic Species*. Vaduz, Liechtenstein: J. Cramer, 1983.

———. "The Sacred Mushroom in Mesoamerica." In *The Ancient Maya and Hallucinogens*, edited by T. Miyanishi, 75–95. Wakayama, Japan: Wakayama University, 1992.

Hart, Carl L. *Drug Use for Grown-Ups: Chasing Liberty in the Land of Fear*. New York: Penguin, 2021.

Haze, Virginia, and K. Mandrake. *The Psilocybin Chef Cookbook*. Toronto: Green Candy Press, 2020.

———. *The Psilocybin Mushroom Bible: The Definitive Guide to Growing and Using Magic Mushrooms*. Toronto: Green Candy Press, 2016.

Hobbs, Christopher. *Medicinal Mushrooms: The Essential Guide*. North Adams, MA: Storey, 2020.

Holland, Julie. *Good Chemistry: The Science of Connection from Soul to Psychedelics*. New York: Harper Wave, 2020.

Huxley, Aldous. *The Doors of Perception*. New York: Harper, 1954.

James, William. *The Varieties of Religious Experience: A Study in Human Nature*. New York: Penguin Classics, 1985.

Janikian, Michelle. *Your Psilocybin Mushroom Companion: An Informative, Easy-to-Use Guide to Understanding Magic Mushrooms*. Berkeley, CA: Ulysses Press, 2019.

Jarnow, Jesse. *Heads: A Biography of Psychedelic America*. Boston: Da Capo Press, 2016.

Lattin, Don. *Changing Our Minds: Psychedelic Sacraments and the New Psychotherapy*. Santa Fe, NM: Synergetic Press, 2017.

———. *God on Psychedelics: Tripping Across the Rubble of Old-Time Religion*. Berkeley, CA: Apocryphile Press, 2023.

———. *The Harvard Psychedelic Club: How Timothy Leary, Ram Dass, Huston Smith, and Andrew Weil Killed the Fifties and Ushered in a New Age for America*. New York: HarperCollins, 2010.

Leary, Timothy, Richard Alpert, and Ralph Metzner. *The Psychedelic Experience: A Manual Based on the Tibetan Book of the Dead*. New York: Kensington, 2022.

Lee, Martin A., and Bruce Shlain. *Acid Dreams: The Complete Social History of LSD: The CIA, the Sixties, and Beyond*. New York: Grove Press, 1994.

Letcher, Andy. *Shroom: A Cultural History of the Magic Mushroom*. London: Faber and Faber, 2006.

Lincoff, Gary, and D. H. Mitchel. *Toxic and Hallucinogenic Mushroom Poisoning: A Handbook for Physicians and Mushroom Hunters*. New York: Van Nostrand Reinhold, 1977.

Martinez, Rebecca. *Whole Medicine: A Guide to Ethics and Harm-Reduction for Psychedelic Therapy and Plant Medicine Communities*. Berkeley, CA: North Atlantic Books, 2024.

McKenna, Dennis. *The Brotherhood of the Screaming Abyss: My Life with Terence McKenna*. Santa Fe, NM: Synergetic Press, 2023.

McKenna, Terence. *Food of the Gods: The Search for the Original Tree of Knowledge—A Radical History of Plants, Drugs, and Human Evolution*. New York: Bantam, 1992.

Menser, Gary P. *Hallucinogenic and Poisonous Mushroom Field Guide*, 3rd ed. Berkeley, CA: Ronin Publishing, 1996.

Muraresku, Brian. *The Immortality Key: The Secret History of the Religion with No Name*. New York: St. Martin's Press, 2020.

Nicholas, L. G., and Kerry Ogamé. *Psilocybin Mushroom Handbook: Easy Indoor & Outdoor Cultivation*. Oakland, CA: Quick American, 2006.

Noconi, Cody. *The Psychedelic History of Mormonism, Magic, and Drugs*. Self-published, 2021.

Nuwer, Rachel. *I Feel Love: MDMA and the Quest for Connection in a Fractured World*. New York: Bloomsbury, 2023.

Orsini, Aaron Paul. *Autism on Acid: How LSD Helped Me Understand, Navigate, Alter & Appreciate My Autistic Perceptions*. Self-published, 2019.

Orsini, Aaron Paul, ed. *Autistic Psychedelic: The Self-Reported Benefits & Challenges of Experiencing LSD, MDMA, Psilocybin & Other Psychedelics as Told by Neurodivergent Adults Navigating ADHD, Alexithymia, Anxiety, Asperger's, Autism, Depression, PTSD, OCD & Other Conditions*. Self-published, 2010.

Oss, O. T., and O. N. Oeric. *Psilocybin: Magic Mushroom Grower's Guide: A Handbook for Psilocybin Enthusiasts*. Oakland, CA: Quick Trading Company, 1993.

Plotkin, Mark J. *Tales of a Shaman's Apprentice: An Ethnobotanist Searches for New Medicines in the Rain Forest*. New York: Penguin, 1994.

Pollan, Michael. *How to Change Your Mind: What the New Science of Psychedelics Teaches Us About Consciousness, Dying, Addiction, Depression, and Transcendence*. New York: Penguin, 2018.

Pollock, Steven H. *Magic Mushroom Cultivation*. San Antonio: Herbal Medicine Research Foundation, 1977.

Rae, Jaden. *Microdosing Guide and Journal: A Complete Step-by-Step Microdose Guide to Safely Source, Track, & Customize Your Medicinal Mushroom Journey*. Self-published, Fresh Karma, 2022.

Rätsch, Christian. *Encyclopedia of Psychoactive Plants: Ethnopharmacology and Its Applications*. Rochester, VT: Park Street Press, 2005.

Richards, William A. *Sacred Knowledge: Psychedelics and Religious Experiences*. New York: Columbia University Press, 2016.

Riedlinger, Thomas J., ed. *The Sacred Mushroom Seeker: Tributes to R. Gordon Wasson*. Rochester, VT: Park Street Press, 1990.

Rogers, Robert. *The Fungal Pharmacy: The Complete Guide to Medicinal Mushrooms & Lichens of North America*. Berkeley, CA: North Atlantic Books, 2011.

Ruck, Carl A. P., and Mark A. Hoffman. *Entheogens, Myth, and Human Consciousness*. Berkeley, CA: Ronin Publishing, 2013.

Sacks, Oliver. *Hallucinations*. New York: Vintage, 2013.

Schultes, Richard Evans, Albert Hofmann, and Christian Rätsch. *Plants of the Gods: Their Sacred, Healing, and Hallucinogenic Powers*. Rochester, VT: Healing Arts Press, 2001.

Shetreat, Maya. *The Master Plant Experience: The Science, Safety, and Sacred Ceremony of Psychedelics*. Self-published, Difference Press, 2023.

Shulman, Lisa M. *Before and After Loss: A Neurologist's Perspective on Loss, Grief, and Our Brain*. Baltimore: Johns Hopkins University Press, 2018.

Smith, Huston. *Cleansing the Doors of Perception: The Religious Significance of Entheogenic Plants and Chemicals*. New York: Penguin, 2000.

Stamets, Paul, ed. *Fantastic Fungi: How Mushrooms Can Heal, Shift Consciousness, and Save the Planet*. San Rafael, CA: Insight Editions, 2019.

Stamets, Paul. *Psilocybin Mushrooms of the World: An Identification Guide*. Berkeley, CA: Ten Speed Press, 1996.

Stamets, Paul, and J. S. Chilton. *The Mushroom Cultivator: A Practical Guide to Growing Mushrooms at Home*. Self-published, Agarikon Press, 1983.

Waldman, Ayelet. *A Really Good Day: How Microdosing Made a Mega Difference in My Mood, My Marriage, and My Life*. New York: Anchor, 2017.

Wasson, R. Gordon. *Soma: Divine Mushroom of Immortality*. New York: Harcourt Brace Jovanovich, 1972.

———. *The Wondrous Mushroom: Mycolatry in Mesoamerica*. New York: McGraw-Hill, 1980.

Wasson, R. Gordon, Albert Hofmann, and Carl A. P. Ruck. *The Road to Eleusis: Unveiling the Secret of the Mysteries*. New York: Harcourt Brace Jovanovich, 1978.

Wasson, R. Gordon, Stella Kramrisch, Jonathan Ott, and Carl A. P. Ruck. *Persephone's Quest: Entheogens and the Origins of Religion*. New Haven, CT: Yale University Press, 1986.

Wittmann, Marc. *Altered States of Consciousness: Experiences Out of Time and Self*. Translated by Philippa Hurd. Cambridge, MA: MIT Press, 2018.

Index

About the Author

Eugenia Bone is the internationally known food and nature writer of eight previous books. She has been nominated for the Colorado Book Award and the James Beard Award, and won a Nautilus Book Award for *Fantastic Fungi Community Cookbook.* Her work has appeared in the *New York Times*, *National Lampoon*, *Saveur*, and *BBC Science Focus*, among many others, as well as the *Wall Street Journal*, where she is a frequent book reviewer.

Bone is a member of the National Association of Science Writers and a former president of the New York Mycological Society. She is on the faculty of the New York Botanical Garden, where she teaches classes on psychedelic mushrooms and mycophagy. She is featured in the documentary *Fantastic Fungi* (2019) and the Netflix children's program *Waffles + Mochi*, produced by Michelle Obama's Higher Ground Productions.